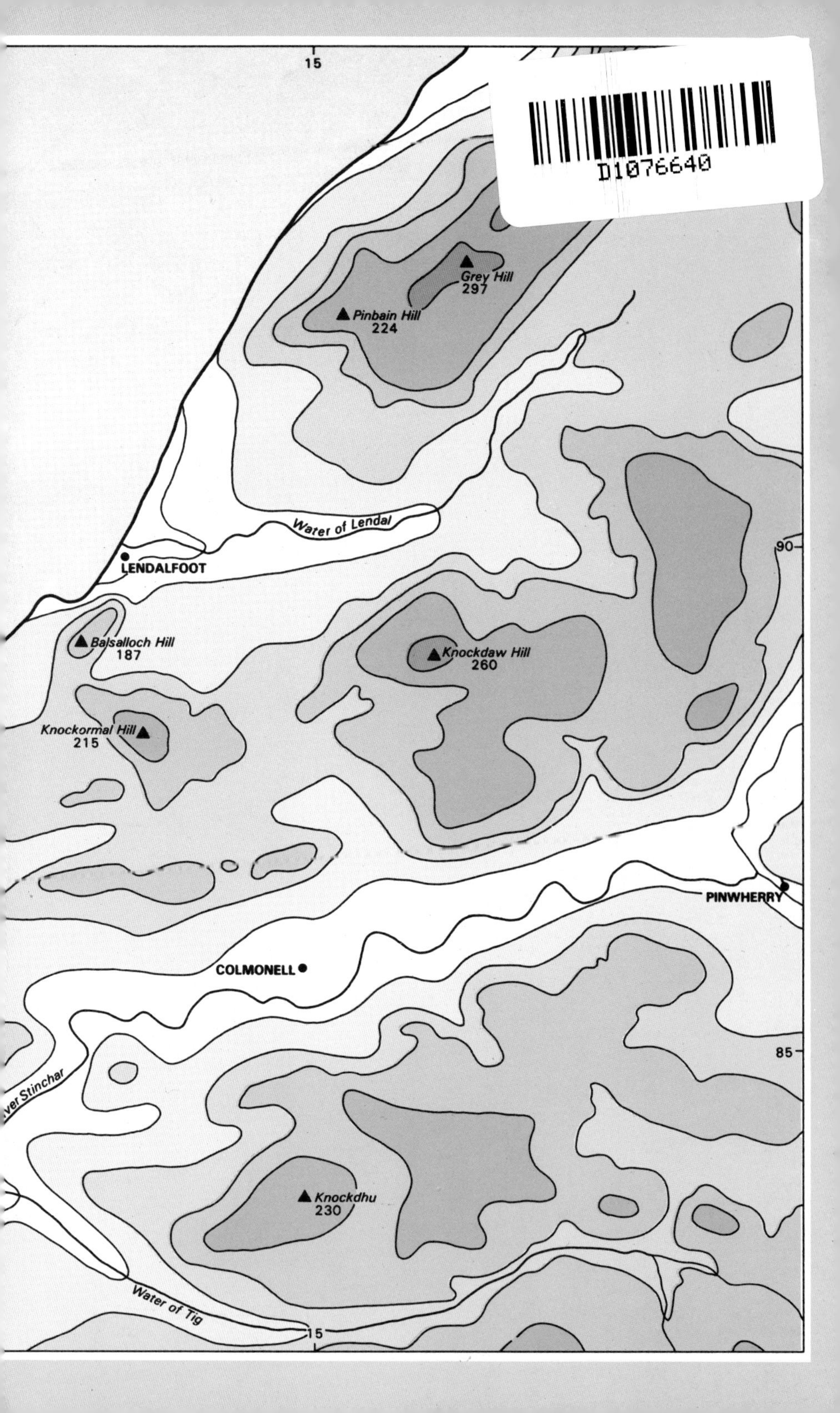
15
D1076640
Grey Hill
297
Pinbain Hill
224
Water of Lendal
LENDALFOOT
90
Balsalloch Hill
187
Knockdaw Hill
260
Knockormal Hill
215
PINWHERRY
COLMONELL
85
ver Stinchar
Knockdhu
230
Water of Tig
15

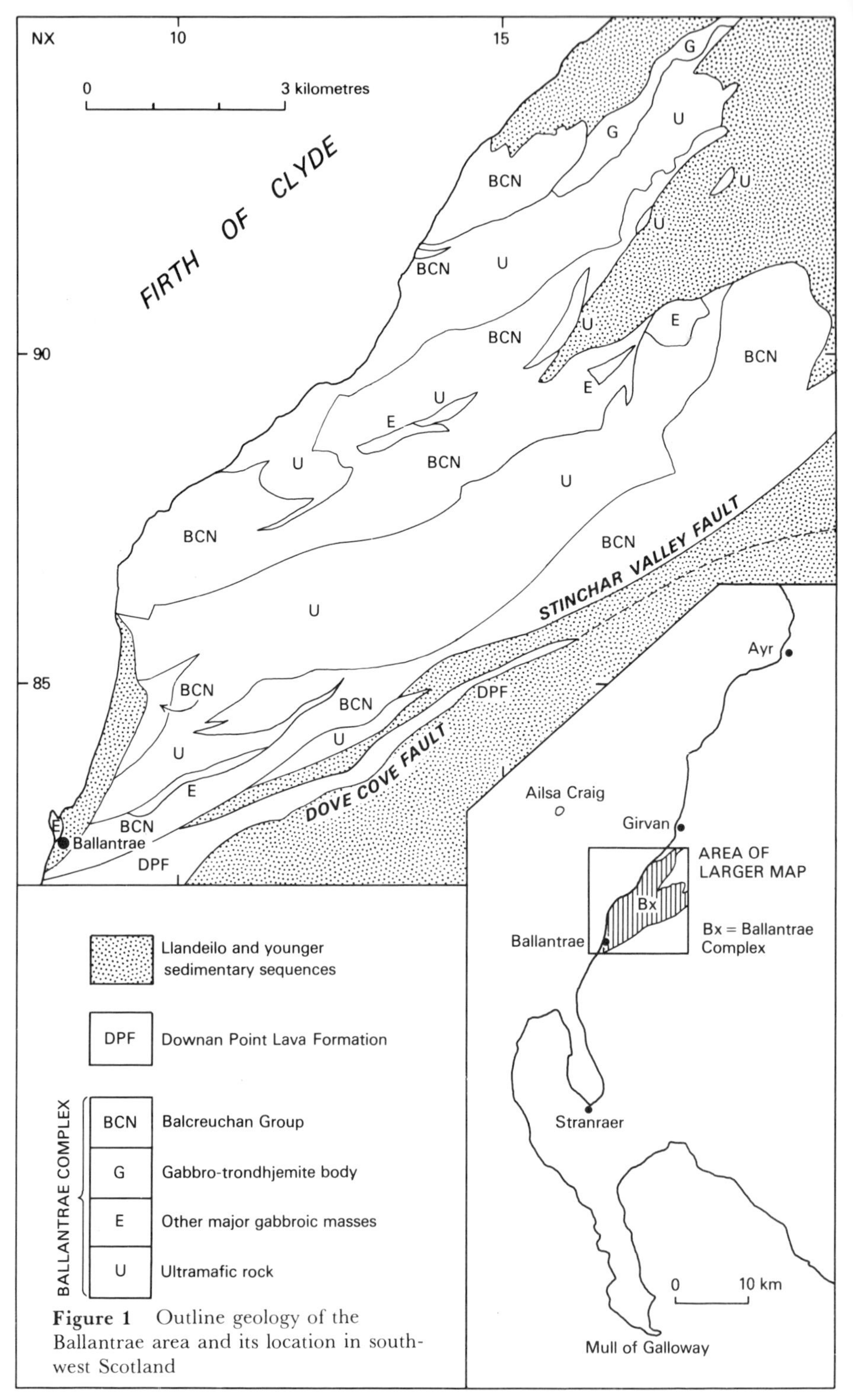

Figure 1 Outline geology of the Ballantrae area and its location in south-west Scotland

Preface

This volume and the complementary 1:25 000 geological map of the Ballantrae area arise from a programme of mineral reconnaissance and mapping carried out by the British Geological Survey on behalf of the Department of Trade and Industry. They represent the first comprehensive re-examination of the district since the 1899 revision by B. N. Peach and J. Horne of the original 1867 survey published at 1 inch to the mile.

In addition to the 1:10 000 and 6″ -scale field mapping, interpretation and laboratory work of the authors and contributors this explanatory text has benefitted from additional information supplied by M. J. Gallagher, A. Davies, A. E. S. Kemp, S. P. Tunnicliff, J. S. Coats, A. G. Gunn, N. M. S. Rock and R. M. Carruthers, all of the British Geological Survey. Dr R. A. Gayer made available material from the collection of the University College, Cardiff, whilst Dr B. J. Bluck of the University of Glasgow allowed access to his unpublished field maps. Diagrams and plates were prepared by the Drawing Office and Photographic Section of the BGS, Edinburgh. The book was typeset in the Book Production Section of the BGS, Keyworth.

F. G. Larminie, OBE
Director

British Geological Survey
Keyworth, Nottingham
6 February 1988

Contents

Illustrations

Tables

Introduction

1

Between Girvan and Ballantrae on the shores of the Firth of Clyde, south-west Scotland, an assemblage of ultramafic and mafic plutonic rocks, basaltic pillow lavas and a range of early Ordovician sedimentary lithologies crops out over about 75 km^2 (Figure 1). The area has long attracted the attention of geologists; Nicol (1844) described the serpentinised ultramafic rocks which reminded Murchison (1851) of those exposed at the Lizard, Cornwall. Geological controversy also began at an early date, initiated by Geikie's (1866) proposal that all of the varied igneous lithologies present were derived by the metamorphism of greywacke, an abundant supply of which was available in the Southern Uplands immediately to the south. The proposed metamorphic origin was reiterated in the Geological Survey Memoir (Geikie and Geikie, 1869) accompanying the first geological map of the area, itself published in 1867. An intrusive origin for the ultramafic rocks was subsequently championed by Bonney (1878) and gradually won acceptance, being adopted by Peach and Horne (1899) in their monumental Geological Survey Memoir. Meanwhile the Arenig age of the sedimentary parts of the assemblage had been established by Lapworth (1889); the age of the serpentinised ultramafic rocks then became the next topic for argument. Anderson (1936) produced further evidence for the intrusion of the serpentinite into the volcano-sedimentary sequence, thus requiring the serpentinite to be Arenig or later. However, almost simultaneously Balsillie (1937) described the serpentinite as unconformably overlain by the volcanic and sedimentary rocks and thus pre-Arenig in age. The subsequent debate between these two became a little acrimonious but remained inconclusive, and the issue was further confused by the suggestion of Bailey and McCallien (1952) that the serpentinite was a submarine lava flow interbedded with the basaltic units. This interpretation did not, however, find general favour. Modern petrological work was initiated by Bloxam and reported in a number of papers, discussed more fully in the appropriate sections of the following text, but it was the development of plate-tectonic models which caused the most significant reassessment of the Girvan-Ballantrae assemblage. The association therein of serpentinite, pillow lavas and chert, the so-called 'Steinmann Trinity' had been noted by Teall (*in* Peach and Horne, 1899) and was further discussed by Bailey and McCallien (1960). It was therefore a natural development for the assemblage to be interpreted as an ophiolite complex (Church and Gayer, 1973), a fragment of oceanic crust tectonically emplaced at a continental margin. This interpretation is now widely accepted but the development of refined geochemical techniques to allow discrimination between basalts of differing petrotectonic origins has allowed some of the volcanic units to be interpreted as polygenetic (e.g. Wilkinson and Cann, 1974). This problem has yet to be fully resolved and will receive more attention in Chapter 3 of this volume.

The plethora of research work on the ultramafic-mafic Girvan-Ballantrae lithological assemblage has resulted in a confused application of descriptive 'stratigraphical' nomenclature. Balsillie (1932) first formally grouped all of the rocks together as the *Ballantrae Igneous Complex*, modified by Williams (1962) to the *Ballantrae Igneous Series*. Bloxam (1954, and in subsequent papers) preferred the *Girvan-Ballantrae Complex* whilst more recently Bluck (1978) has used the simpler version of the *Ballantrae Complex.* Following their interpretation Church and Gayer (1973) wrote of the *Ballantrae ophiolite*, an expression also favoured by Bluck (1982) and further strengthened by Stone and Rushton (1983) as the *Ballantrae ophiolite complex*.

Meanwhile Greig (1971), Wilkinson and Cann (1974) and Spray and Williams (1980) all followed Balsillie's usage. With the serpentinite interpreted as a submarine lava flow Bailey and McCallien (1952) placed all of the ultramafic rocks and basaltic lavas into the *Ballantrae Volcanic Series* and, despite general disagreement with their proposals Ingham (1978) modified the term to the *Ballantrae Volcanic Group*, and still included within it all of the disparate mafic and ultramafic plutonic rocks, the extrusive basalts and the Arenig sedimentary sequences.

The historical consensus favours Ballantrae rather than Girvan-Ballantrae as the geographical designation of what the majority of authors agree is a complex. In view of the wide range of rock types present, including sedimentary sequences, any descriptive term for the entire assemblage had best avoid the use of igneous or volcanic. Similarly the likely polygenetic nature of much of the volcanic succession requires caution in the application of the term ophiolite. The simplest solution therefore would seem to be a *Ballantrae Complex* defined to include all of the mafic and ultramafic plutonic rocks together with the xenoliths and tectonic inclusions therein, and the interbedded volcanic pillow lavas and Arenig sedimentary sequences with which they are geographically associated. To the south the Complex is bounded by the Stinchar Valley Fault, to the north and east it is unconformably overlain by Llanvirn and younger sedimentary rocks, and to the west its submarine continuation is probably terminated either by faulting against, or by an unconformity below, Permian strata occupying a basin in the Firth of Clyde. This is the meaning of the term *Ballantrae Complex* as used in the following text and on the accompanying 1:25 000 geological map. For lithostratigraphical convenience the Arenig volcano-sedimentary parts of the Ballantrae Complex can be further defined as a Group, preferably avoiding the confusing repetition of the prefix Ballantrae, and this course is followed in Chapter 3 of this account with the establishment of the Balcreuchan Group. Further subdivision into locally significant formations has been attempted where appropriate.

The sedimentary sequences of Llanvirn and younger age which unconformably overlie the Ballantrae Complex have a less confused history of research. Peach and Horne (1899) and Lapworth (1882) provided a sound base to the understanding of the succession, whilst aspects of the sedimentology and style of deposition were clarified by Henderson (1935) in a remarkably perceptive piece of work for his time, and by Kuenen (1953). A modern stratigraphical framework was provided by Williams (1962) and Ingham (1978) and is broadly followed in this account. South of the Stinchar Valley Fault Williams was also able to refine Peach and Horne's original stratigraphy but this area of the map remains the least satisfactory in that respect.

Both the Ballantrae Complex and the overlying younger sedimentary sequence are generally well exposed on the coast south of Girvan. Inland the level of exposure may be moderately high on some of the hills but over much of the area it is uniformly low This is particularly true for the ultramafic rocks, which have been preferentially eroded in such a way that a broad variation in topography is controlled by the outcrop pattern of different lithologies. The dominant structural trend throughout the Complex is NE–SW, with major faults separating elongate zones of alternately ultramafic-mafic plutonic and volcano-sedimentary lithological assemblages, so producing a NE–SW alignment of topographical relief. This effect was accentuated by glacial scouring directed west or south-west, now evinced by abundant crag and tail features, roches moutonnées and ice-smoothed knolls.

A particularly good example of the differing responses to erosion of the principal lithologies within the Ballantrae Complex is provided by the coastal section. There, outcrops of the volcano-sedimentary sequences are usually terminated at their seaward end by steep and rocky cliffs whereas the serpentinite outcrop is marked by an extensive raised beach backed by cliffs of glacial till. Coastal serpentinite exposure is thus restricted to the shoreline platform. Elsewhere in the ultramafic outcrop tectonic inclusions, xenoliths or minor intrusions are often preferentially exposed to form prominent 'knockers' with serpentinite exposed only along their margins. It is these that have been moulded into the best examples of glacial crags and roches moutonnées.

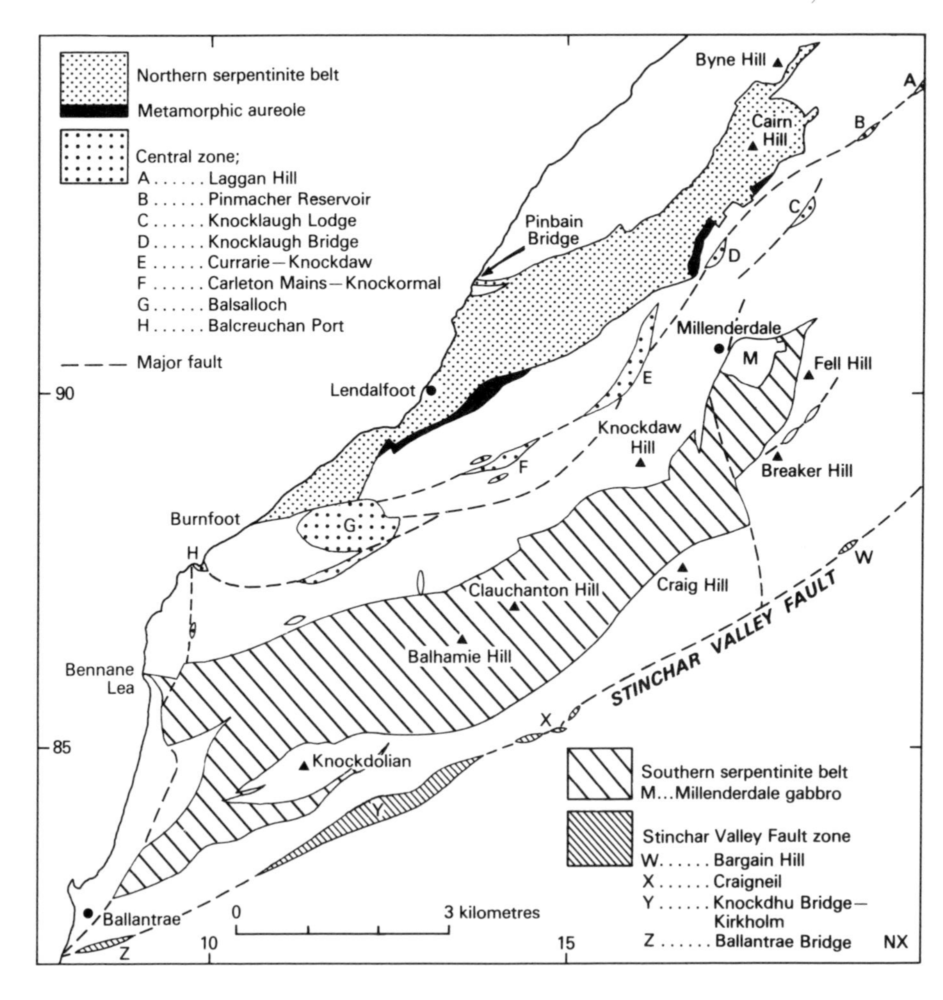

Figure 2 Principal outcrops of ultramafic rock within the Ballantrae Complex

2 Ultramafic and associated metamorphic and mafic igneous rocks of the Ballantrae Complex

The ultramafic rocks of the Ballantrae Complex are predominantly dunites and harzburgites which have been affected by variable, but usually extensive, serpentinisation. The resultant serpentinite invariably has a variegated appearance which depends largely on the proportion of three textural components: a generally black and green mottled groundmass, brown bastite pseudomorphs, and pale green lustrous veins. Examination by optical, electron probe and XRF analytical techniques of specimens, collected at outcrop in all parts of the Complex and obtained as core during drilling operations (Stone and others, 1984), has allowed their comprehensive classification in terms of the textural categories defined by Wicks and Whittaker (1977) and by Wicks and others (1977).

The great majority of the serpentinite is pseudomorphic after olivine or orthopyroxene with an asymmetric mesh developed from the former rather more commonly than a symmetric form. Mesh rims are usually of α-serpentine, probably lizardite, with only very rare development of γ-serpentine (?antigorite) and are generally bipartite, the bipartite:tripartite ratio being about 10:1. Relict olivine survives in the centre of some mesh cells, but hourglass texture of α-serpentine is more common. Fibrous and isotropic mesh centres, often containing γ-serpentine, were also widely observed. In a few specimens the pseudomorphic mesh merges to form curtain texture of α-serpentine, rarely accompanied by γ-serpentine. Bastite pseudomorphs after orthopyroxene, generally enstatite, are widespread and are invariably of γ-serpentine. All of these textural variations can be accommodated by the ideal models for serpentinisation proposed by Wicks and others (1977), and reflect original differences in the olivine structure of the protolith rather than variations in the serpentinisation process. The great preponderance of pseudomorphic mesh textures suggests that most of the serpentinisation was governed by a non-shearing, falling or constant temperature regime (Wicks and Whittaker, 1977), probably in the range 100–300°C (Wenner and Taylor, 1971). Large quantities of water are required for the transformation of peridotite to serpentinite: stable isotope studies (e.g. Magaritz and Taylor, 1974; Barnes and others, 1972) indicate that meteoric or connate water is responsible, introduced from the surrounding country rock. Calcium metasomatism, probably associated with the serpentinisation process, has affected clinopyroxene in particular to produce grossularite (and more rarely andradite) aggregates occasionally intergrown with prehnite.

Non-pseudomorphic textures were recognised in a small number of the specimens examined, with both interlocking and interpenetrating forms present. The serpentine grains, invariably of the γ variety, are anhedral and, even when more elongate in the interpenetrative texture, are randomly orientated to produce a massive rather than a foliated rock. Antigorite flame or thorn texture is well developed locally. Most of the non-pseudomorphic serpentine texture was seen to be replacing pseudomorphic forms and so may indicate a second, localised and higher-temperature phase of serpentinisation or metasomatism, converting predominantly α-serpentine/lizardite into γ-serpentine/antigorite.

Numerous serpentine veins cut all of the other textures described and are of both cross- and slip-fibre type; the cross-fibre veins are the largest (ranging up to 5 mm across) and the most abundant. All of the cross-fibre veins examined were of γ-serpentine but both γ- and α-serpentine were observed in the slip-fibre veins.

The serpentinisation processes were penecontemporaneous phenomena, probably late Arenig or earliest Llanvirn in age, and the lithological textures they produced are equally apparent in all parts of the Ballantrae Complex. However, significant differences in the ultramafic protolith exist between different parts of the Complex, and these will be systematically described in the succeeding section. The locations of the ultramafic outcrops within the Ballantrae Complex are shown in Figure 2. Lithological terminology used is summarised in Figure 3 following the scheme proposed by Coleman (1977, p.26).

Northern serpentinite

The outcrop of the northern serpentinite forms a belt 1–1.5 km wide extending north-east from Burnfoot [NX 1053 8812] to Byne Hill [NX 1810 9487], a distance of about 11 km. Within this area the topography is subdued and undulating except to the south and east of Cairn Hill [NX 1765 9350] where serpentinite crops out in a series of crags. Along the coast, the relatively low resistance of the serpentinite to weathering and erosion has facilitated the creation of a gently dipping boulder-strewn foreshore platform backed by prominent raised beaches and a fossil cliff-line cut in boulder clay (well developed around Lendalfoot [NX 1320 9006]).

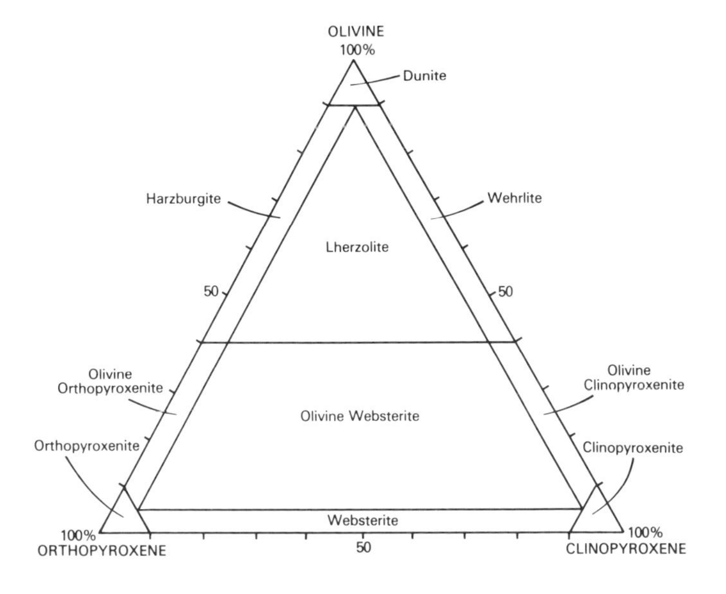

Figure 3 Descriptive terminology for ultramafic rocks in terms of their relative content of olivine, orthopyroxene and clinopyroxene

Figure 4 Bastite pseudomorphs after orthopyroxene contained in a symmetrical mesh of serpentinised olivine. Crossed polars, ×30 S 68556, PMS 461

Harzburgite

Lithologically, the outcrop consists principally of serpentinised harzburgite (Figure 4), an olivine-orthopyroxene rock in which olivine is characteristically replaced by serpentine minerals, and orthopyroxene (originally enstatite) is replaced either by bastite, a bronze-coloured lustrous variety of serpentine, or more rarely by pale green actinolitic amphibole. Relics of both olivine and enstatite are preserved in the centres of incompletely altered pseudomorphs. Spinel is a ubiquitous accessory mineral in the ultramafic rock; at one locality in the raised cliff-line north of Pinbain Bridge [NX

1386 9152] it becomes a major component. There, millimetre-size chromite grains are concentrated in a zone 4–5 m wide of serpentinised olivine chromitite. This shows faint, steeply inclined banding or rare, crudely graded layering caused by variation in chromite content from about 30% to 95%. Locally the rock is brecciated such that angular clasts of chromitite are contained in a foliated and serpentinised dunite matrix. In thin section (S 77252), the rounded and embayed spinel crystals and rare pseudomorphs after enstatite, which may poikilitically enclose serpentinised olivine crystals, attest to an origin for the harzburgite by crystal settling and accumulation.

In most exposures the serpentinised harzburgite is a massive rock with a saccharoidal texture on freshly broken surfaces, but it is smooth and waxen on polished and striated joints and some weathered surfaces. It is generally deep green and typically weathers orange-brown, but may be a distinctive yellow-green with deep green fractures. Reticulate veining by pale green, or white-weathering fibrous serpentine is also characteristic. Locally, a pink or red hue is caused by dissemination of hematite within the serpentine mesh, e.g. on inland crags about 1 km north of Laigh Knocklaugh, and the foreshore at Games Loup near Burnfoot. The exposure there is unusual in that it shows an interfingering of green and red serpentinite, with the green variety apparently veining the red (Figure 5). This relationship suggests that the colour differences are related to lithological and/or chemical variation within the original harzburgites, but the serpentinites now differ petrologically only in hematite content and in a greater concentration of bastite pseudomorphs in the green variety. Elsewhere,

Figure 5 Veined relationship between petrologically similar serpentinites. Chrysotile veinlets lie parallel to the contacts. Games Loup. D 3345

weathering effects apart, colour diversity within the serpentinite may be caused by variation in the degree of serpentinisation, in particular the distribution of secondary magnetite exsolved during alteration of olivine (Wicks and Whittaker, 1977).

Balsillie (1932) first described a fine foliation within the serpentinite, which he considered had a metamorphic (pre-Arenig) origin although Jones (1977) subsequently attributed this texture to crystal accumulation. The foliation is a mineralogical banding, ranging from 1 mm to 5 cm in width and caused by alternating enstatite and olivine-rich layers; the more resistant pyroxenous bands form narrow ridges separated by depressions eroded in the softer olivine-rich rock. Although volumetrically minor, spinel crystals are preferentially associated with the olivine-rich layers (Jones, 1977). The foliation is only patchily preserved. It occurs at a few isolated localities on the foreshore and hill slopes between Burnfoot and Carleton Burn being best seen at Carleton Fishery [NX 1253 8944]. However, it is more widely developed farther north-east between Currarie and Cairn Hill. There it has a general north-north-westerly strike swinging round to north-east within a zone 100 m wide close to the faulted south-eastern margin of the ultramafic rock. The dip of the foliation varies randomly between 45° and vertical.

Pyroxenite

The northern serpentinite outcrop also contains a range of mineralogically and texturally diverse pyroxenites which, being harder than the serpentinite and relatively resistant to weathering, tend to form pale greyish, coarsely crystalline knolls protruding above the surrounding serpentinite. The pyroxenites vary principally in the relative proportions of diopside and enstatite, and the presence or absence of significant accessory phases, notably olivine (Figure 3). In comparison with the advanced alteration state of the enclosing serpentinite they are characteristically little altered: diopside may show marginal or patchy replacement by actinolite or chlorite, and enstatite is often partially pseudomorphed by bastite, but olivine is invariably completely serpentinised.

Owing to the poor exposure, the field relationships of the pyroxenites are not well known. Jones (1977) has described websterite interbanded with harzburgite or dunite, which he collectively termed 'banded cumulate lherzolite', and it is likely that other pyroxenites occur in comparable banded or lensoid form. By contrast, beach exposures on the south side of Bonney's Dyke [NX 1350 9108] reveal a network of vein-like, coarse clinopyroxenite apparently injected into the serpentinised harzburgite, against which, however, the veins are not chilled.

Textural variations and their implication

Texturally, the pyroxenites can be divided into two types. Those which crop out in the western and north-western parts of the serpentinite belt are coarse- to very coarse-grained lithologies formed of interlocking platy pyroxene crystals up to 3 cm across, some showing twinning (e.g. S 69941, 69945). They have allotriomorphic-granular textures, sometimes poikilitic in part, which were probably created by normal magmatic crystallisation (Pike and Schwarzman, 1977). Pyroxenites which crop out farther east, within about 0.5 km of the south-eastern margin of the ultramafic outcrop, tend to be fine grained and characterised by heteroblastic textures, comprising large strained pyroxene porphyroclasts dispersed in a finer recrystallised matrix. The texture is variably developed and ranges from (in the coarsest rocks) marginal granulation and minimal recrystallisation, to ragged porphyroclasts 'floating' in a matrix of finer pyroxene polyhedra (neoblasts) with curved or straight crystal boundaries commonly meeting at 120° (e.g. S 69946). Some of these rocks also show a coarse foliation, seen in thin section as a preferred, nematoblastic mineral orientation, confirming these lithologies as metamorphic tectonites formed by subsolidus recrystallisation under stress at mantle pressures and temperatures (cf. Nicolas and others, 1971; Basu, 1977; Pike and Schwarzman, 1977).

The separation into tectonised and magmatic textures is not immediately apparent in the associated harzburgites, largely on account of the masking effects of pervasive serpentinisation. Thin sections simply show pseudomorphic mesh textures characteristic of olivine replacement, traversed by a variable number of slip- and cross-fibre serpentine veins. In places, all traces of the original texture have been destroyed and replaced by a dense mass of anastomosing fibrous serpentine. This is caused by low-temperature recrystallisation of serpentine during late-stage, presumably high-level brittle faulting and should not be confused (cf. Jones, 1977) with mantle tectonite fabrics. However, close examination of thin sections suggests that a broad, bipartite division can be made which has significance regarding primary textures. Serpentinite from western and northern exposures shows a coarse mesh texture, with former crystal boundaries interlocked with large rounded and embayed spinel crystals and bastite pseudomorphs which, together with some poikilitic texture is suggestive of crystal accumulation (Jones, 1977). By contrast, within the area of tectonised pyroxenites, the serpentinites have noticably finer mesh texture suggesting replacement of smaller olivine primocrysts, although original grain boundaries are seldom recognisable. Further, the associated spinel grains are typically intensely fractured or even disrupted, with some showing a tendancy to 'clumping' along grain boundaries. These serpentinites are interpreted as tectonised harzburgites for the following reasons:

1 Olivine has recrystallised under stress to form a finer-grained polycrystalline aggregate.

2 Spinel grains have been pushed aside by recrystallisation of the surrounding ferromagnesian minerals with the formation of clumped 'hollyleaf' textures.

3 In some exposures a crude foliation is defined by the parallel orientation of bastite pseudomorphs. Although in size and form the bastite crystals are comparable with bastites seen in the non-tectonised harzburgite, enstatite recrystallises less readily than olivine and it may have been protected by the enclosing, more abundant olivine crystals, which preferentially absorbed the deformative effects.

Marginal features of the serpentinite outcrop

The tectonised harzburgites and pyroxenites are apparently confined to a narrow zone along the south-eastern margin of the serpentinite outcrop. Within this zone, the relict cumulate banding undergoes a progressive reorientation. Moreover, an unusual assemblage of metamorphic lithologies generally lies between the serpentinite and the adjacent lavas and sedimentary rocks, cropping out in three main areas (Figure 2): on the seaward slopes of Balsalloch Hill and Carleton Hill, in the Lendal Water above and below Straid Bridge [NX 1390 9005], and on the flanks of the elevated moorland north-west and north of Laigh Knocklaugh. The metamorphic rocks are divisible into an upper (north-western) unit of amphibolites and a lower (south-eastern) unit of slaty epidote schists. Because of incomplete exposure, contact relationships are often uncertain, although the upper contact with serpentinite seems invariably to be faulted. Other contacts within and at the base of the metamorphic zone are locally gradational. However, faults are common and are probably responsible for the variation in both the relative thickness of the two metamorphic units and in their combined thickness which ranges from 200 m at Straid Bridge to 20 m in stream exposures 1.1 km north-north-east of Laigh Knocklaugh.

The metamorphic rocks were originally considered to be a continuous thermal aureole formed either along the margins of a hot peridotite intrusion and modified by shearing (Peach and Horne, 1899; Anderson, 1936) or during cooler, dilational emplacement of the serpentinite (Bloxam, 1955), or as fragments of country rock incorporated by the rising magma and concentrated along its margins (Lewis, 1975). Recent, more detailed work (Jones, 1977; Spray and

Williams, 1980; Treloar and others, 1980) lays emphasis on the pronounced pressure-, as well as temperature-, inversion which occurs across the comparatively narrow zone and is incompatible with a static origin as a conventional aureole. As originally suggested by Church and Gayer (1973), the zone was likened to a dynamothermal metamorphic aureole, formed by progressive 'welding' onto the sole of a hot and extensive ultramafic body during thrusting initiated, at considerable depth, within oceanic lithosphere and ending with obduction onto continental crust.

By comparison with the complex history associated with the southern margin, the northern margin of the serpentinite outcrop appears to be a simple steeply dipping normal fault or, posssibly, a north-westerly-translating oversteepened thrust (Williams, 1959). Contacts are exposed at only two localities: on the foreshore 70 m south-west and 160 m north of Pinbain Bridge [NX 1375 9142]. There, the adjacent sedimentary rocks have clearly undergone shearing and bedding is locally truncated. The serpentinite is brecciated, veined and locally replaced by carbonate within a zone 2 m wide parallel to the contact. The presence of serpentine veins in spilitic lava at the more northerly contact (Anderson, 1936) demonstrates that at least some serpentinisation may have occurred following emplacement of the lava (Bailey and McCallien, 1954, 1957), but it is also explicable as back-injection caused by frictional heating generated during faulting or thrusting.

Bonney's Dyke

A striking pegmatitic gabbro up to 5 m wide known as Bonney's Dyke (Balsillie, 1932, p.116) protrudes from the ultramafic rocks on the foreshore about 1 km north of Lendalfoot [NX 1351 9108]. It has a west-north-west trend but has an arcuate outcrop owing to the cumulative effects of numerous minor sinistral faults. The gabbro is in places very coarse grained (Figure 6) and shows no sign of chilling against the surrounding serpentinite and pyroxenite, which were probably hot when the gabbro was emplaced. It was thus broadly coeval with the ultramafic rocks as a differentiate of the same magma. Comparable intrusions crop out on the foreshore about 250 m to the north [NX 1376 9136] and at the foot of the fossil cliff line to the east [NX 1363 9111], but these examples are smaller and are poorly exposed.

Bonney's Dyke has been described in detail by Bonney (1878), Heddle (1878) and Jones (1977). It is a pale-coloured rock formed almost exclusively of plagioclase and clinopyroxene, now largely replaced by a secondary rodingite assemblage of pectolite, prehnite, pumpellyite, actinolite, chlorite, garnet, and occasional serpentine after original olivine. Three types of marginal contact are

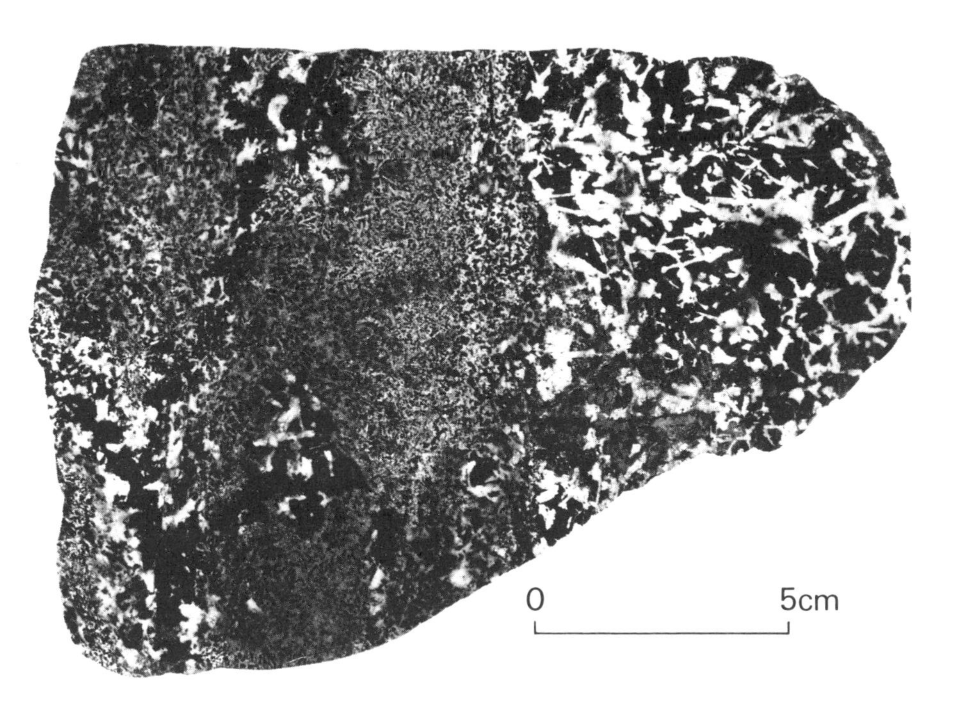

Figure 6 Grain size variation within coarse gabbro pegmatite. Bonney's Dyke MNS 4007

displayed: sharp contacts against serpentinite (including serpentinite enclaves); less well defined margins abutting the coarse pyroxenite; and sheared contacts that are fine grained, flinty and particularly intensely rodingitised. Within the general thesis of a layered mafic/ultramafic complex developed by Jones (1977), the gabbro was regarded as a steeply dipping crystal cumulate with at least three graded units. However, these are only developed on a single face of a large loose block and cannot be traced laterally. They are atypical of the intrusion as a whole, which characteristically shows marked grain-size variation from medium to very coarse in a random manner similar to other pegmatitic occurrences (Figure 6).

The metamorphic aureole adjacent to the northern serpentinite

The assemblage of schistose rocks and amphibolites at the south-eastern margin of the northern serpentinite belt (Figure 2) has been compared to the inverted metamorphic aureoles at the base of many classical ophiolites (Spray and Williams, 1980; Treloar and others, 1980). It is believed to have formed during the thrusting of a hot peridotite body over upper oceanic crust.

Structurally, the outcrops are dominated by a 'main phase' of slaty schistosity (epidote schists) or fine gneissic foliation (amphibolites) with a patchily developed mineral-aggregate lineation. These features dip or plunge steeply to the north-north-west. In addition, the schist contains tight, asymmetrical, essentially reclined interfolial folds closing south-west and north-east; these structures are refolded by tight, angular, straight-limbed folds with associated axial-planar crenulation cleavage. Such a history of polyphase deformation restricted to a narrow band of rocks is characteristic of a mylonite zone (Treloar and others, 1980).

The epidote-schists (e.g. S 67580) are slaty rocks characterised by flaser banding: streaked epidote augen and granular boudins up to a few millimetres in diameter are dispersed in finer-grained dark green layers of actinolitic hornblende, chlorite and albite. Light yellow-green layers of fine-grained epidote and sphene are also present together with occasional metasedimentary quartz-albite-epidote-muscovite-chlorite schists and pods of recrystallised carbonate. By comparison, the amphibolites (e.g. S 69972) are coarser-grained, dark grey, very hard rocks largely composed of plagioclase and green or brown hornblende. Spray and Williams (1980) noted the occurrence of better foliated garnet amphibolites (e.g. S 69968) with brown hornblende and pyroxene in exposures closest to the serpentinite, but this simple distribution is often confused by interdigitation related to faulting. At one locality [NX 1767 9287] this has led to imbrication of serpentinite and metamorphic lithologies.

Treloar and others (1980) described lenses of hornblende-bearing garnet metapyroxenite from a few isolated localities close to the amphibolites but separated from them by a narrow band of sheared serpentinite. Although the metapyroxenites could form the highest-grade part of the metamorphic assemblage, they are believed to represent segregations within the ultramafic sequence which were fortuitously juxtaposed close to the base of the ultramafic outcrop and affected by processes occurring within the mylonitic sole, albeit in the early stages (cf. Spray, 1982). The metapyroxenite has been dated at 505 ± 11 Ma (Sm-Nd; Hamilton and others, 1984). Texturally and in their mineralogy and mineral chemistry, the garnet metapyroxenites indicate recrystallisation under upper granulite facies or mantle conditions ($T = 900 \pm 70°C$, $P = 10 - 15$ kbar: Treloar and others, 1980). By contrast, at least part of the amphibolite sequence may have formed at $T = 850°C$ and $P = 7$ kbar (Spray and Williams, 1980), and the epidote schists formed under very much lower-grade conditions corresponding to the low epidote amphibolite and/or greenschist facies.

Central zone

Within a zone of anastomosing faults trending north-east to south-west approximately midway between the two main serpentinite belts, there lie a number of small, lenticular outcrops of ultramafic rock (Figure 2). Blue amphibole-bearing foliated basalts are also present within this zone, emphasising its importance as a fundamental shear across the Ballantrae Complex. Those ultramafic inliers in the north-east of the zone are the least well exposed, and two, at Pinmacher Reservoir [NX 1933 9376] and Knocklaugh Lodge [NX 1824 9211], were first located by their associated magnetic anomalies (Powell, 1978; Carruthers, 1980). The inlier at Knocklaugh Lodge was subsequently proved by drilling (Stone and others, 1984) to consist principally of dark green, serpentinised harzburgite in which bronze-brown bastite grains commonly range up to 3 mm dimeter. Locally these form elongated aggregates which define a weak foliation, possibly an original feature of the protolith, and in one or two specimens the proportion of bastite increases to a degree which suggests olivine-orthopyroxenite as the original lithology. The majority of the rocks examined are cut by abundant pale green serpentinite veins; shearing is pervasive throughout the borehole section; and the fault-bounded margins of the outcrop are associated with extensive silicification and carbonatisation. From the very limited exposed evidence available the Knocklaugh Lodge serpentinite would seem to be typical of the small bodies of ultramafic rock in the north-east of the central zone (A to E on Figure 2). These crop out in the vicinity of Laggan Hill [NX 2015 9440], Pinmacher Reservoir [NX 1933 9376], Knocklaugh Lodge [NX 1824 9211], Knocklaugh Bridge [NX 1727 9180], and in a lenticular strip between Currarie Farm [NX 1622 9094] and Knockdaw [NX 1506 8960].

Approximately 1 km south-west of Knockdaw another fault-bounded slice of ultramafic rock crops out midway between Carleton Mains and Knockormal (around NX 1425 8910; F on Figure 2). This outcrop contains what is probably the best known of all the unusual lithologies in the Ballantrae Complex: a hornblende-garnet clinopyroxenite (S 71750) described as eclogite by its discoverer, Balsillie (1937), and discussed in some detail by Bloxam and Allen (1960). Both of these accounts were based on material collected from a large loose boulder, but a very similar pyroxenite lithology (S 59417) was exposed in situ nearby in a temporary trench. It contains tschermakitic hornblende, fassaitic clinopyroxene and almandine-rich pyrope garnet, and has been interpreted (Smellie and Stone, 1984) as a segregation within mantle harzburgite. Serpentinised harzburgite, wehrlite and clinopyroxenite are exposed close by, although the field relationships are not clear. The clinopyroxenite is locally brecciated, probably as a result of major faulting at the outcrop margins. The sheared basalts at these margins contain a sparse development of blue amphibole (Bloxam and Allen, 1960) but the mineral chemistry of the so-called eclogite (more properly a garnet clinopyroxenite) is not compatible with an origin within a high-pressure/high-temperature metamorphic belt (Fettes, 1978). Rather, the chemistry supports a segregation origin in the lowermost crust or upper mantle at possible pressures of 10–13 kbar and temperatures around 900–950°C (Smellie and Stone, 1984). This garnet clinopyroxenite segregation is the oldest known part of the Ballantrae Complex with a Sm-Nd age of 576 ± 32 Ma reported by Hamilton and others (1984).

The south-western part of the central zone contains the largest outcrop of ultramafic rock outside the two main serpentinite belts (G on Figure 2). This outcrop occupies about 1 km^2 around Balsalloch [NX 1175 8826] and at its northern margin is in faulted contact with the northern serpentinite belt. The southern margin is also faulted and is marked by an extensive zone of silicification and carbonatisation. In contrast, the eastern and western margins appear to dip moderately westward and are interpreted as early thrusts cut by the later faulting. Serpentinised harzburgite is the dominant rock type present but the bastites after orthopyroxene are generally small (less than 2 mm) and rarely make up more than 20% of the rock. Relict olivine is preserved in some specimens but serpentinisation is generally complete with the fine-grained serpentine mesh cut by abundant thin serpentine veins. Hematite veining is also common and the distribution of fine-grained hematite around the mesh imparts a red colour to much of the ultramafic rock.

In the extreme south-west of the central zone, on the coast at Balcreuchan Port [NX 0985 8755], serpentinite crops out in a small fault-defined body well exposed below high-water mark (H on Figure 2). The eastern margin of the outcrop is a major north-south fault adjacent to which the serpentinite is pervasively sheared and has been affected by patchy silicification and carbonatisation. The western and southern margin of the serpentinite is interpreted as a south-westerly-dipping thrust but, since downthrow on the major north–south fault is probably to the west, this is unlikely to be a continuation of the marginal structures of the Balsalloch body. Dunite was probably the protolith for most of the Balcreuchan Port serpentinite which now shows a fine mesh, often with calcite grains in the mesh centres. Veining by serpentine, calcite and prehnite is common, and rare hornblende veins also occur. Most of the serpentinite has been affected by Ca metasomatism with the development of fine grossular aggregates in places. This may indicate the original presence of wehrlite since the grossular is most likely to have replaced clinopyroxene.

South of the central zone proper but north of the southern serpentinite belt two north–south-trending faults introduce thin (less than 30 m across) slices of sheared serpentinite into the volcanic lava and breccia sequence at Meikle Bennane [NX 0974 8666] and to the west of Lochton Hill [NX 1295 8736]. In both outcrops the ultramafic lithology is a serpentinised fine-grained harzburgite cut by abundant serpentinite veins. The presence of these slices of serpentinite may indicate the occurrence of more serpentinite at depth beneath a wider area of the volcanic succession's outcrop.

Southern serpentinite

The southern serpentinite outcrop occupies about 17 km^2 disposed in a linear belt up to 2 km broad extending from the coast at Bennane Lea [NX 092 859] north-eastwards for approximately 10 km to Fell Hill [NX 183 905]. It is separated at the northern and southern margins from the adjacent volcanic sequences by steeply inclined faults. Exposure in the western part of the outcrop is very poor with faulting of the ultramafic rocks against both Arenig basalts and Permian sandstones limiting coastal exposure of serpentinite to the Bennane Lea area itself. In the central, inland, part of the serpentinite belt, the upland areas of Balhamie, Clauchanton and Breaker hills provide more extensive exposures, the south-west slope of Balhamie Hill [NX 131 862] being probably the best exposed section of ultramafic rock (a coarse bastite serpentinite) in the whole Complex. Farther northeast the serpentinite outcrop is increasingly obscured by forestation.

Harzburgite

Serpentinised harzburgite is volumetrically the most abundant constituent of the southern serpentinite belt with bronze-brown bastite pseudomorphs after orthopyroxene, probably enstatite, ranging up to 2 cm in diameter enclosed in serpentinised olivine (Figure 4). As noted by Jones (1977) the bastite pseudomorphs in the southern serpentinite belt tend to be both larger and more numerous than those seen in the northern outcrop but still seldom account for more than 20% of the rock.

Some variation in the nature of the protolith to the southern serpentinite is indicated by a change in the MgO/SiO_2 ratio of the ultramafic rocks (Coleman, 1977, pp.100–101). North-east of Clauchanton Hill all of the specimens analysed had MgO/SiO_2 ratios in the range 0.90 to 0.96, which is compatible with a harzburgite protolith. By contrast specimens collected between Clauchanton Hill and the sea have MgO/SiO_2 ratios in the range 1.08 to 1.17, which is indicative of a protolith intermediate between harzburgite and dunite. It therefore seems possible that the overall proportion of orthopyroxene in the southern serpentinite belt decreases westward.

Whatever the protolith the serpentinite produced is a generally massive, deep olive-green or dark bluish green rock sporadically reddened by a high content of hematite. Hematite may also form discrete veins but is more frequently dispersed around the serpentine mesh in association with magnetite from which it has developed. A fine granular texture may be apparent on freshly broken surfaces but the rock often splits along smooth and waxy, anastomosing joint planes many of which are slickensided. The coarser bastite serpentinite (e.g. S 68556) is a more massive rock which, despite ubiquitous fine kink bands in the bastite pseudomorphs, is hard and comparatively dense; so much so that at Balhamie this lithology has been extensively quarried for building stone. The quarry itself [NX 1348 8598] is now overgrown and difficult of access but the rock exploited may be seen well exposed in the adjacent hillside and as blocks in the neighbouring field walls. The characteristic pitting of weathered surfaces of serpentinised harzburgite, which is due to the preferential erosion of the bastite grains, is well illustrated at this locality.

Cumulate lithologies

Dunite and more rarely wehrlite (Figure 7) may occur with the harzburgite throughout the ultramafic outcrop. The proportion of wehrlite would by higher if all of the secondary

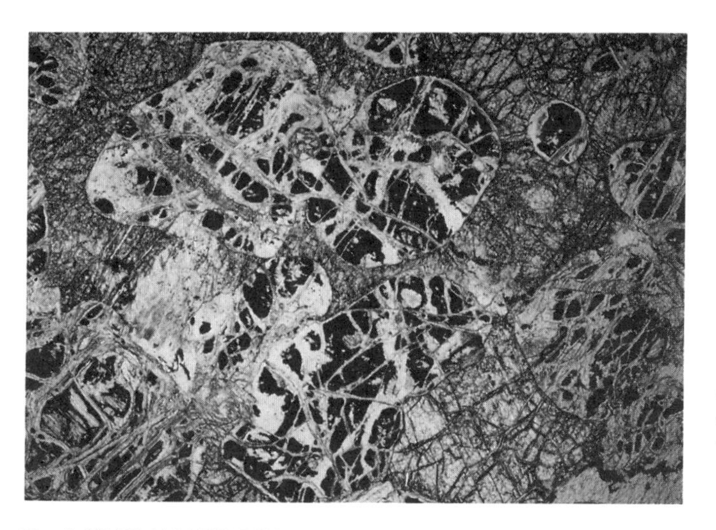

Figure 7 Olivine grains poikilitically enclosed by clinopyroxene in a wehrlite. Crossed polars, ×30 S 68574, PMS 463

grossularite aggregates in the serpentinite are taken as evidence for pre-existing clinopyroxene. Relationships between the rock types are usually obscure but an interdigitation between the dunite and the harzburgite can be established locally, and more rarely the dunite seems to form dyke-like masses within harzburgite. Elsewhere a tectonic relationship is suggested, with pods of unfoliated wehrlite or dunite contained within an area of sheared bastite serpentinite after harzburgite. There is a tendency for the proportion of wehrlite to increase towards the southern margin of

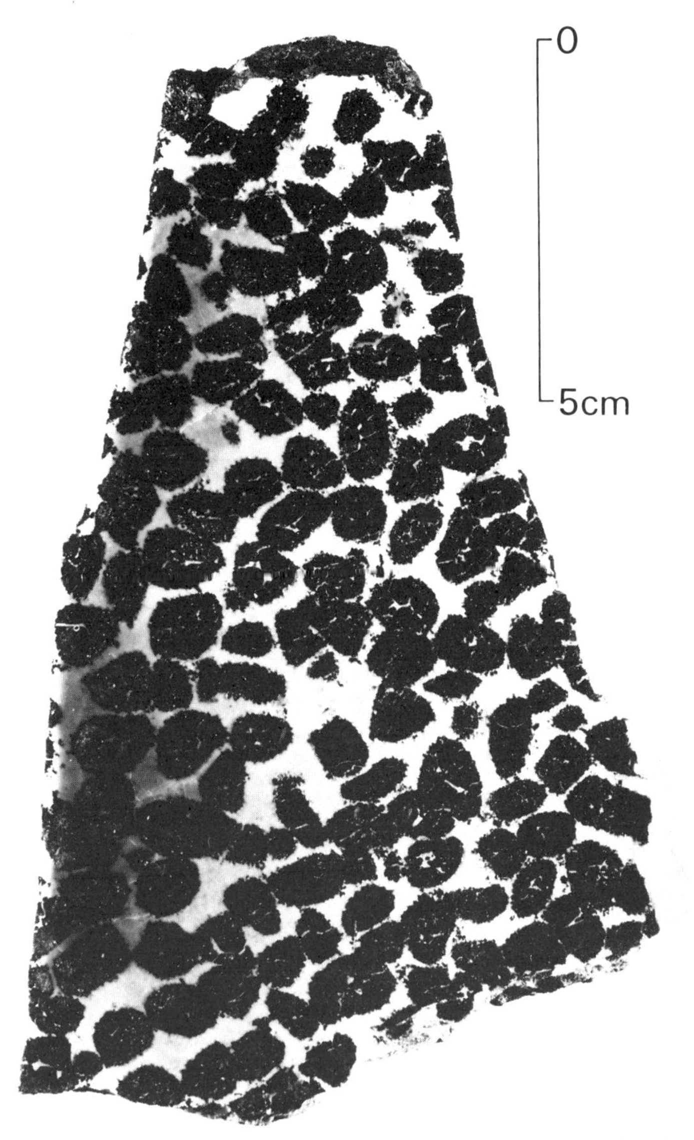

Figure 8 Chrome spinel nodules in serpentinised dunite. Poundland Burn. MNS 3899

the ultramafic belt, and this rock type, together with other more unusual lithologies, is also volumetrically more significant in the area between Knockdaw Hill, Craig Hill and Breaker Hill (Figure 2). In this sector of the ultramafic outcrop layered relationships are evident between harzburgite, dunite, wehrlite, rare clinopyroxenite and olivine gabbro, the dunite in one exposure [NX 1698 8820] containing spectacular nodules of chrome spinel (Figure 8). These range up to 1 cm in diameter and in some specimens (e.g. S 77245) form almost 90% of the rock. This occurrence was first described by Teall (in Peach and Horne, 1899, p.469) as veins of massive picotite, and indeed the chrome spinel-rich zones have a marked dyke-like appearance. However, the whole assemblage of lithologies was compared by Jones (1977) to part of the transition zone seen between the ultramafic and gabbroic units of typical ophiolite sequences and interpreted as a series of cumulate cyclic units. Balsillie (1932, p.112) had earlier described the wehrlites as layered ultramafic cumulates.

The harzburgites and dunites, as elsewhere in the Complex, form a continuous series with bastite pseudomorphs after orthopyroxene comprising over 50% of some specimens of the former but decreasing to traces or a complete absence in the latter. Chrome spinel and magnetite are ubiquitous accessories in both lithologies. Wehrlite, well seen in layered contact with harzburgite [NX 1678 8868] (S 77254 and 77258) and at several other localities in the same area] is typically made up of olivine phenocrysts, variably anhedral to euhedral and ranging up to 5 mm diameter, contained within large plates of clinopyroxene, usually augite (Figure 7). The proportion of clinopyroxene varies in the specimens examined from about 10% to almost 60%, with a mode in the 25–30% range, but the poikilitic relationship is everywhere preserved, the olivine, together with embayed chrome spinel grains, being set in a continuous web of augite. Locally the augite has been replaced by grossularite and more rarely by brown hornblende. Only one exposure of probable clinopyroxenite was noted in this area but that was considerably altered with anhedral grains of diopside (at least 50% of the rock) set in an aggregate of prehnite and grossularite. Serpentine and brown hornblende are minor accessories and it seems likely that this lithology was originally composed principally of diopsidic clinopyroxene.

In the ophiolitic transition zone (*s.s.*) deduced by Jones (1977) the base is marked by the appearance of plagioclase as an intercumulus phase; a phenomenon producing olivine-gabbros which crop out at several localities between Knockdaw Hill and Breaker Hill. In one place the olivine gabbro (S 77255) has a sharp contact with coarse bastite-serpentinite, a relationship which may be intrusive although there are no signs of chilling. Elsewhere detailed relationships are obscure and the olivine-gabbro exposures simply have a close spatial association with dunite and wehrlite. In the gabbros large anhedral or subhedral serpentinised olivines are set in a mosaic of labradorite plagioclase. Small anhedral augite grains are also present but are extensively altered to brown hornblende, which is itself cut by late serpentine veins. Accessory phlogopite has been reported (Jones, 1977) in some of the gabbros in this area. The common accessory opaque minerals chrome spinel and magnetite, are accompanied in some of the gabbros by traces of ilmenite, bravoite, bornite and chalcopyrite.

The similarity of the ultramafic lithological assemblage in the Knockdaw Hill-Craig Hill-Breaker Hill area to the transition zone sequence of ophiolites seems compelling evidence for the interpretation of the Ballantrae example as such a sequence (Jones, 1977; cf. Church and Gayer, 1973). The implication of this is to make the ultramafic rock of this area the structurally highest sequence seen in either of the serpentinite belts; in particular the ultramafic rocks stand in marked contrast to the tectonised harzburgite and pyroxenite of the northern belt, which represent a structurally much deeper level within the ophiolite ultramafic pseudostratigraphy. However, it is not possible to proceed beyond this generalisation since cyclic cumulate units are common throughout the layered sequence and transition zone of ophiolite ultramafic sections. Numerous examples have been described (e.g. Jackson and others, 1975) illustrating considerable variation between sequences, but all show an overall upward progression from olivine-chromite cumulates through alternating olivine-clinopyroxene and plagioclase-olivine-clinopyroxene cumulates into olivine-gabbros. The top of individual cyclic units has frequently been removed which, together with other evidence for penecontemporaneous disruption of the cumulate layering, emphasises the dynamic conditions under which crystal differentiation occurred.

Textural and lithological variation at the southern margin

Further variation in the nature of the ultramafic protolith to the southern serpentinite belt became apparent through detailed examination of core recovered during drilling operations at Knockdolian Knowes [NX 1280 8553] summarised by Stone and others (1984). There an ultramafic breccia was recognised (S 77248, 77249) which contained discrete fragments of pseudomorphic mesh-textured serpentinite, probably after dunite, either in direct contact with bastite aggregates or separated by serpentinite veins from them. The bastite aggregates probably represent original orthopyroxenites. The correct brecciation-serpentinisation sequence is difficult to deduce, a problem aggravated by the superimposition of a second, non-pseudomorphic serpentinisation texture. However, the most likely explanation is early tectonic brecciation and in this respect the position of the section close to the Stinchar Valley Fault zone may indicate an early initiation of that structure. The other interesting feature of the borehole section is the possible presence both of orthopyroxenite fragments in the breccias and of serpentinised olivine orthopyroxenite (S 77250). Pyroxenites are not common in the southern serpentinite belt, in marked contrast to their abundance in parts of the northern outcrop. The southern belt wehrlites (e.g. S 68574) rarely contain more than 50% clinopyroxene (usually augite), and only one, highly altered, probable clinopyroxenite was observed. Similarly the orthopyroxene content of the harzburgite rarely exceeds 30% (Jones, 1977, reports a range of 24% – 32% obtained by point counting).

Apart from the Knockdolian Knowes borehole examples, only one other specimen studied from the area of the southern belt is composed dominantly of pyroxene: this lithology (S 72038) is found in a narrow fault-bounded sliver of ultramafic rock which crops out in the bed of the River Stinchar to the south-east of Knockdolian [NX 1198 8469]. Both orthopyroxene (probably bronzite) and clinopyroxene (augite) are present in approximately equal proportions and both enclose small anhedral olivine grains. The olivine is fresh and unaltered for the most part but some serpentine pseudomorphs are evident, and actinolitic amphibole, patchily distributed throughout the specimen and in places intergrown with prehnite, replaces both of the pyroxenes and occasionally the olivine. Anhedral brown garnet appears to have a primary relationship with the pyroxenes but since electron microprobe analysis proved only andradite and grossularite a secondary origin seems more likely. Accessory opaque minerals include chrome spinel, magnetite and chalcopyrite. This lithology is a lherzolite but is an isolated occurrence in an ultramafic outcrop otherwise formed of dunite and harzburgite within which it is probably a discrete segregation. However, the position of the outcrop within a major fault zone does allow an alternative explanation involving the faulting into the sequence of material derived from deeper in the ultramafic succession.

Ophicarbonate

The faulted margins of the southern serpentinite belt, and other major structures cutting the ultramafic rocks therein, are spatially associated with the replacement of the mafic minerals by carbonate (usually dolomite) and quartz to produce an 'ophicarbonate' lithology. These rocks are variably pink, grey and creamy brown, and range in texture from fine grained and homogeneous to examples with an apparently contorted fabric giving them a pseudo-gneissose or brecciated appearance. Several successive phases of veining are evident, the earliest usually introducing calcite and dolomite and the later ones quartz. All stages in the production of the 'ophicarbonates' can be seen, from serpentinite cut by a few thin carbonate veins to massive carbonate-quartz rock devoid of mafic minerals. The textural variation in the carbonate-quartz 'ophicarbonate' end products ranges from a mass of coarse-grained carbonate veins and vugs to a fine-grained quartzite containing vugs of chalcedony cut by coarser-grained quartz veins. However, the ubiquitous presence of accessory chrome spinel is a good guide to the origin of the 'ophicarbonates' as original ultramafic rocks. Chlorite is a common accessory mineral, especially in the more foliated examples, which become in effect quartz-chlorite schists invariably veined by carbonate and more quartz. Similar 'ophicarbonate' rocks are associated with many other serpentinite bodies elsewhere in the world, in particular with their faulted margins, and are generally thought to be the result of alteration of the ultramafic protolith by seawater circulating in fault zones at relatively small depths (e.g. Cortesogno and others, 1981). However, in an approximately equal-volume serpentinisation of olivine by the addition of water only, large quantities of MgO and SiO_2 must be removed in solution. At the same time calcium may be released from pyroxenes, an important side-effect in the specific example of the Ballantrae Complex where calcium-metasomatism is widespread. It therefore seems likely that the formation of the 'ophicarbonate' lithologies was associated with either one or both of the serpentinisation episodes and was probably contemporaneous with the extensive calcium-metasomatism discussed more fully in Chapter 4 in the section on rodingitisation. In this respect it may be significant that the most extensive 'ophicarbonate' development is seen in the south-western part of the southern serpentinite belt, and the best evidence for a second phase of serpentinisation comes from the same general area adjacent to the Stinchar Valley Fault zone. Other good examples crop out in the Balsalloch serpentinite body of the central zone as previously described.

Basic igneous xenoliths in the southern serpentinite

A prominent feature of the southern serpentinite belt is the large number of included masses of hornblende or pyroxene hornfels and granulite (*s.l.*), the whole assemblage being considered by Tyrrell (1909) as original plutonic rocks caught up as xenoliths and thermally metamorphosed by a mass of hot peridotite. Subsequent investigators preferred an intrusive origin: Balsillie (1937) proposed tectonic disruption of hot dolerite and gabbro dykes followed by regional metamorphism, whereas Beveridge (1950) and Bailey and McCallien (1952) related the thermal metamorphism to repeated intrusion of dolerite in the same area. Bloxam (1955) followed Tyrrell in preferring the incorporation of dolerite xenoliths in an intrusive peridotite and described the granoblastic-textured rocks as beerbachites, a term which has since been generally used. Jelinek and others (1980) proposed a more sophisticated history for the beerbachites. Their model requires that gabbro and dolerite dykes generated at an active oceanic ridge were not rapidly transported away from the high heat-flow environment. As a possible mechanism transform faulting was invoked to maintain the relative positions of the dolerite-dyke and gabbro complex and the oceanic ridge. However, all of these studies concentrated on the granoblastic beerbachites and overlooked their close spatial association with the texturally dissimilar hornblende-hornfels lithologies which had been subjected to a much lower grade of metamorphism. If this factor is considered the xenolith assemblage suggests either polyphase dyke intrusion into a cooling peridotite or fragments of a disrupted metamorphic aureole. In either case the xenoliths have been subjected to a subsequent retrogressive phase of metamorphism, possibly during the obduction of the ophiolitic ultramafic sequence.

All of the xenoliths are dark brown to dark green rocks, generally uniformly fine grained but occasionally with porphyritic or porphyroclastic textures caused by large (5 mm) rectangular grains of feldspar. Despite the marked textural variations all have a similar mineralogy, with plagioclase, generally andesine, the principal mineral, accompanied by hornblende and augitic pyroxene. The plagioclase is rarely fresh, frequently being pseudomorphed by fine-grained aggregates of prehnite and less commonly of grossularite. The amounts of hornblende and pyroxene in the specimens are inversely proportional to each other and since there is good microscopic textural evidence of hornblende replacing pyroxene, it is probable that much, perhaps all, of the hornblende has formed at the expense of the pyroxene (Jelinek and others, 1980). The hornblende hornfelses only show incipient recrystallisation with a relict igneous texture visible, whereas the beerbachites are almost completely recrystallised rocks. Remarkably, rocks of such significantly different textures and, by implication metamorphic grades, occur within tens of metres of each other, and sometimes within the same xenolithic mass.

Hornblende hornfels

In the southern serpentinite outcrop hornblende-hornfels xenoliths are best exposed in the vicinities of Craig,

Clauchanton and Balhamie hills (Figure 2). The original igneous texture, generally ophitic and occasionally variolitic, is well preserved, and some specimens are coarse enough to qualify as metadolerites, with laths of groundmass plagioclase between 0.3 and 1 mm long (e.g. S 73587). Except for two olivine-bearing specimens none of them show any direct evidence of the recrystallisation of primary pyroxene and only a few specimens show significant recrystallisation of plagioclase. The principal effect of thermal metamorphism is the partial replacement of primary pyroxene by hornblende; the uralitisation process. In some specimens even this replacement is only patchily developed so that areas of unaltered dolerite and areas of hornblende hornfels are transitional, the one into the other.

Beerbachite

The texture of the beerbachites, also well exposed in the Craig Hill and Balhamie Hill areas (e.g. S 73589, 73590), is in striking contrast to that of the adjacent hornblende hornfelses. Their fine-grained (0.1 to 0.3 mm), equidimensional and polygonal, granoblastic texture (Figure 9) demonstrates, apart from rare porphyroclasts of plagioclase and pyroxene (e.g. S 73596), the almost complete recrystallisation of the original igneous minerals. They are clearly high-grade thermally metamorphosed rock (Bloxam, 1955, 1980; Jelinek and others, 1980). Plagioclase, generally andesine, makes up between 30 and 70 per cent of the rock, and in most of the specimens brown hornblende (ferroan pargasite) is the most abundant mafic mineral, with pyroxene (salite or diopside) present in minor amounts or rarely absent. A number of specimens collected from xenoliths north of Craig Hill have in addition accessory amounts of olivine and hypersthene; a characteristic mineral assemblage of the pyroxene-hornfels facies. The beerbachites nearly always show a heterogeneous distribution of the mafic minerals on the one hand, and of plagioclase on the other hand, such that in most specimens the mafic minerals and plagioclase are concentrated in separate lenticular domains a few millimetres long and less than 1 mm wide (Figure 9). The olivine-bearing beerbachites show a mineral layering in which plagioclase layers about 1 mm thick alternate with mafic mineral layers, some of which are dominated by olivine. Beerbachites with a homogeneous texture do occur but these contain generally only, or mainly, hornblende associated with the plagioclase.

The Millenderdale gabbro-dolerite suite

Non-foliated, foliated and flasered gabbros (M on Figure 2) variably metamorphosed to hornfels and beerbachite and cross-cut by numerous beerbachite dykes crop out as a series of prominent rocky knolls about 600 m east of Millenderdale [NX 1715 9055] (Peach and Horne, 1899, p.477; Balsillie, 1932; Jelinek and others, 1980). They are thought to represent a single, very large xenolith. The gabbros are pale coloured, medium- to very coarse-grained rocks (e.g. S 71249) that locally show a steep, north-west-dipping to subvertical foliation predominantly striking north-north-east. Bluck and others (1980) report a K-Ar age for a flaser gabbro of 487 ± 8 Ma. No contacts with the surrounding serpentinite are exposed, but an abrupt change to doleritic lithologies takes place to the west across a line parallel to the main foliation. That some of the foliation must be a primary feature due to igneous crystallisation differentiation is indicated by the variation in pyroxene compositions between different bands, and the localised presence of olivine. The gabbros are texturally variable, ranging from those produced by normal igneous crystallisation to granoblastic and nematoblastic textures developed under high-grade metamorphic conditions.

The beerbachite dyke rocks are medium-grained, dark grey dolerites with a saccharoidal texture locally weathering to a rust-brown colour. Although some of the dykes are reported to have chilled margins, most do not, indicating that the host rock was still hot at the time of dyke injection. The dykes have a very wide range of orientations and no dominant trend is evident. They were probably not emplaced simultaneously, as evidenced by the cross-cutting relationships observed. Some dykes follow a tortuous anastomosing path and may enclose masses of foliated gabbro. Those injected at an early stage are now finely foliated (S 71054), some showing augen structure. They are invariably orientated parallel to the predominant north-north-east foliation in the host gabbro, with which the dyke margins are gradational, and they never intrude the commoner, non-foliated beerbachites. A few of the beerbachite dykes have a very faint internal foliation that bears no simple relation to the main gabbro foliation, and they may cut other weakly or non-foliated beerbachites. Most commonly, the non-foliated dykes (S 71055, 71058) sharply cross-cut the gabbro foliation, at a variety of angles up to 90°. However, concordant relationships are also present and in places the gabbro foliation bends into conformity at the dyke wall, a feature suggesting that the gabbro was in a plastic state at the time of dyke injection.

Chemically, the gabbros and beerbachites probably form a genetically related, ocean-floor tholeiite suite with the more evolved gabbros interpreted as the upper differentiates of a layered assemblage which crystallised in a magma chamber in oceanic crust. The beerbachites have been described as forming part of a sheeted dyke complex (Jelinek and others, 1980) although they typically form only 20–25 per cent of

Figure 9 Typical beerbachite polygonal, granoblastic texture with heterogeneous distribution of mafic minerals and plagioclase S 27363
Top: Polygonal granoblastic texture developed between feldspar and brown amphibole. Plane polarised light, ×40 PMS 469
Bottom: Development of mafic and leucocratic domains. Plane polarised light, ×20 PMS 470

the exposed rock. Similar but less well developed examples of this beerbachite association also crop out in a fault-bounded slice about 2 km south-west of Millenderdale on the northern slopes of Knockdaw Hill [NX 166 895].

Metamorphic history

The textural evidence indicates that much of the hornblende in the beerbachites has replaced granoblastic pyroxene grains with all stages of replacement from incipient to nearly complete sometimes seen in the area of a single thin section. Elsewhere pyroxene, both primary and secondary in the hornblende hornfels lithologies and also, significantly, in the neighbouring pyroxene-bearing ultramafic rocks, has been similarly replaced by hornblende. Since its crystallisation is clearly ubiquitous it is logical to refer the formation of the hornblende to a late, hydrous and retrogressive phase of metamorphism. The earlier metamorphic event recorded in the xenoliths involving recrystallisation of pyroxene and plagioclase is consistent with the progressive thermal metamorphism of dolerites and gabbros. Thus the xenoliths, which show all stages of this recrystallisation from incipient to complete, can most simply be envisaged as members of the same thermal aureole. It is possible that such an aureole could be formed in the way described by Jelinek and others (1980) whereby there was '. . . movement along a transform fault so that dykes emplaced laterally across such a fault, from a spreading ridge, are kept adjacent to the high heat flow environment of the ridge'. However, this would produce the same high metamorphic grade over a wide area and is difficult to reconcile with polyphase dyke intrusion. The close spatial association of xenoliths of markedly different grade may therefore favour an explanation in which the base of a dyke and gabbro complex, part of an ophiolite sequence, is intruded at depth by a hot, peridotitic crystal-laden 'magma' (cf. Tyrrell, 1909; Bloxam, 1955). This event may be dated by the K-Ar age of 487 ± 8 Ma obtained by Bluck and others (1980) from a gabbro at Millenderdale and the K-Ar age of 475 ± 8 Ma given for a gabbro north of Colmonell by Harris and others (1965). The relatively narrow thermal aureole thus produced, an outward progression from pyroxene beerbachite to lower-grade hornfels merging with the unmetamorphosed country rock, was then disrupted by the further, probably tectonic, movement of the partially cooled peridotite mass, possibly in association with obduction of the ophiolite. At this stage, and with the likely addition of a hydrous phase, hornblende replaced pyroxene throughout the suite at the same time as fragments of the aureole, and unmetamorphosed country rock, were incorporated as xenoliths. Subsequent minor phenomena such as the formation of biotite and the extensive development of prehnite and grossularite at the expense of feldspar have a separate paragenesis, with the latter probably related to the Ca-metasomatism seen as an associate of serpentinisation of the peridotite host.

Stinchar Valley Fault zone

The Stinchar Valley Fault marks the south-eastern limit of the Ballantrae Complex *sensu stricto*. At its south-westerly, seaward end it is close to the faulted margin of the southern serpentinite belt but the two structural lineaments diverge slightly north-eastward. The zone has a likely polyphase structural history and includes elements of both the Ballantrae Complex and its upper Ordovician cover sequence as lenses contained in an anastomosing fault network. One large and several smaller bodies of ultramafic rock crop out therein.

South of Bargain Hill [NX 1890 8777, W on Figure 2] a thin sliver of ophicarbonate is the most north-easterly evidence of ultramafic rock in the Stinchar Valley Fault zone. There is no further exposure for about 4 km south-westward until, in the Craigneil area [NX 1438 8524, X on Figure 2], three small, discrete outcrops of ultramafic lithologies have some unusual characteristics. The most easterly of these, exposed in Pyet Glen, consists of highly sheared, completely serpentinised, fine-grained harzburgite. However, despite the shearing the rock remains fairly massive, in marked contrast to the other two outcrops exposed a little farther west. In the Craigneil Burn small blocks of serpentinite are contained in a soft and friable greenish-grey clay, apparently the product of weathering of sheared serpentinite since the trace of a fabric is preserved in the clay. The bastite serpentinite blocks are angular in this exposure, but 500 m westward in a small, unnamed stream a similar soft, greenish grey, serpentinous clay contains apparently well rounded pebbles of bastite serpentinite and more angular blocks of gabbro pegmatite. The gabbro-pegmatite blocks are often contained as tectonic inclusions in the southern serpentinite belt but their association here with rounded pebbles indicates that this ultramafic outcrop, and possibly that exposed in the Craigneil Burn, may be original serpentinite conglomerates. If so, the conglomerates may be part of the upper Ordovician Barr Group, the cover sequence to the Ballantrae Complex from which Williams (1962) has described similar lithologies, or its lateral equivalent the Tappins Group. Conglomeratic facies of both of these groups, but without serpentinite pebbles, crop out elsewhere in the Stinchar Valley Fault zone.

The largest mass of ultramafic rock in the zone (Y on Figure 2) crops out between Knockdhu Bridge [NX 1334 8454] and Kirkholm [NX 1122 8380]. Typically the lithology is a mottled, dark green, bastite serpentinite after harzburgite but in some exposures an advanced second phase of serpentinisation has given the rock a nodular appearance. Pods of pseudomorphic serpentinite produced during the first serpentinisation episode are now contained as relicts within a 'host' of non-pseudomorphic serpentinite produced during the second episode. Veining by calcite, serpentine and hematite is common. No more ultramafic rock is exposed in the Stinchar Valley Fault zone between Kirkholm and the sea but red serpentinite has been reported near Ballantrae (Z on Figure 2) from borehole core recovered during site investigation work [NX 0846 8222] adjacent to the River Stinchar (BGS Archives, Edinburgh).

Economic mineralisation in the ultramafic rocks

The only evidence for historical attempts at exploitation of economic mineralisation in the Ballantrae Complex is the abandoned trial workings in the north of the area (Figure 10) near Byne Hill [NX 1807 9438] and Balkeachy [NX 1829 9336]. These investigated sparse copper mineralisation in shear zones adjacent to bodies of ultramafic rock but the genesis of that mineralisation is uncertain. Elsewhere the ultramafic rocks of the Complex contain a range of opaque, accessory ore minerals. These include a variety of Cu-Ni minerals present only as traces and chrome spinel as a ubiquitous accessory. The usually fractured and embayed chrome spinel grains rarely exceed 1 mm diameter except at two localities (Figure 10) where both the size and abundance of the spinels are dramatically increased.

Figure 10 Location map of known mineralisation within the ultramafic rocks of the Ballantrae Complex

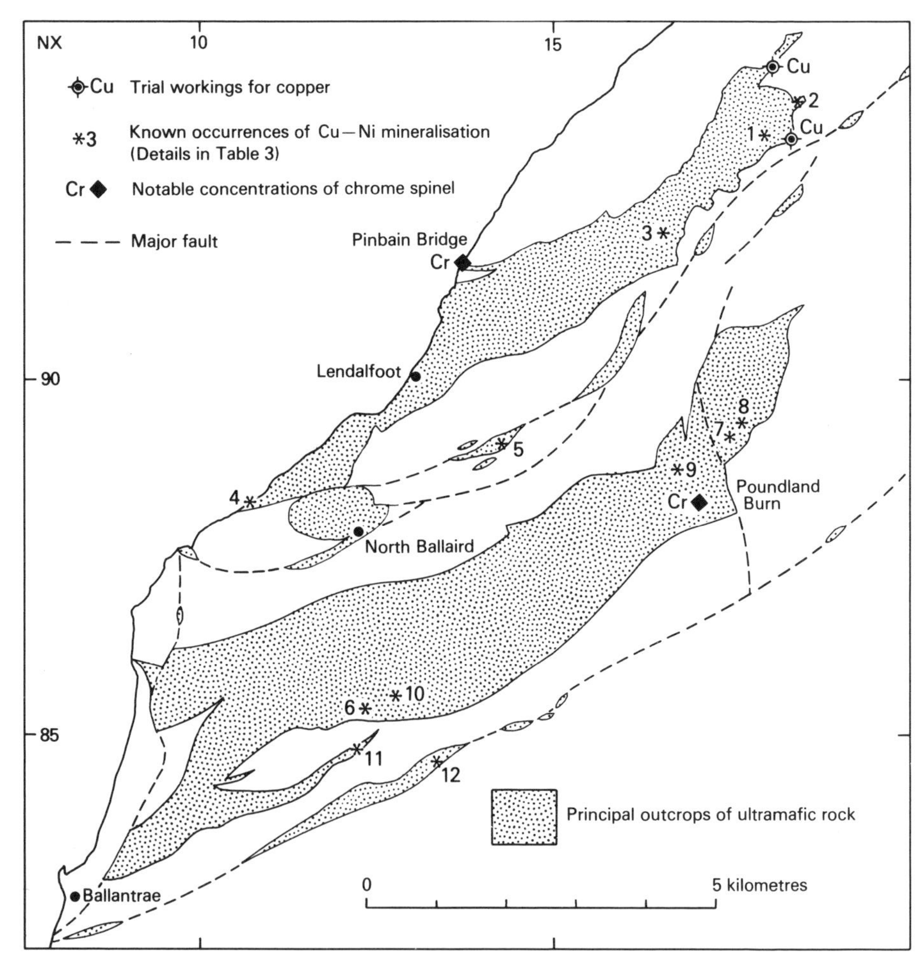

Sulphide mineralisation

The distribution of Cu-Ni mineralisation traces is summarised in Figure 10 and Table 1. Pyrrhotite was only identified in the northern serpentinite belt but generally the southern serpentinite is the more mineralised of the two major outcrops, albeit still at very low levels, in terms of both Ni and Cu. In both ultramafic outcrops there is an appar-

Table 1 Known occurrences of Cu-Ni mineralisation in the ultramafic rocks of the Ballantrae Complex

Mineral	Protolith of host lithology*	Specimen details+	Location=
Native Cu	Olivine orthopyroxenite	Borehole at 63.75 m S 77250	10
	Harzburgite	Borehole at 108.0 m S 77251	10
Covellite	Wehrlite	Inclusions in magnetite S 77256	7
Chalcocite	Olivine orthopyroxenite	Borehole at 63.75 m S 77250	10
	Harzburgite	Borehole at 108.0 m S 77251	10
Chalcopyrite	Hornblende clinopyroxenite segregation in harzburgite	Intergrown with bornite S 77246	4
	Olivine gabbro	Inclusion in magnetite S 77257	8
	Lherzolite segregation in harzburgite	S 72038	11
Bornite	Clinopyroxenite	Intergrown with pyrrhotite S 77243	1
	Hornblende clinopyroxenite segregation in harzburgite	Intergrown with chalcopyrite S 77246	4
	Olivine gabbro	Inclusion in magnetite S 77257	8
Millerite	Harzburgite	S 77253 in x-fibre serpentine vein	6
		S 77247	12
	Harzburgite and Dunite	Borehole at 46–47 m S 77248, 77249	
Bravoite	Wehrlite	Inclusions in magnetite S 72256	7
	Olivine gabbro	S 77255	9
Pyrrhotite	Clinopyroxenite	S 77243 intergrown with bornite	1
		S 69945	3
	Harzburgite	Intergrown with goethite in carbonate veins S 77241	
	Wehrlite	Rimmed by goethite S 67606	5

* Now extensively serpentinised

+ Numbers refer to specimens housed in the collection of the BGS, Edinburgh

= See Figure 10

ently increased likelihood of mineralisation in pyroxene-rich rocks, a trend which is more marked since these lithologies form the volumetric minority of the ultramafic sequence. The exception to this is millerite, which was only observed in harzburgite and dunite. Apart from those minerals summarised in Table 1 there is also an unconfirmed report of pentlandite contained in a chloritic vein cutting the Poundland Burn (Figure 10) nodular picotite lithology (British Geological Survey, unpublished data).

The 'ophicarbonate' lithologies formed by the carbonatisation and silicification of the peridotites contain a variety of accessory ore minerals together with ubiquitous pyrite and marcasite. The most highly mineralised are those cropping out in the vicinity of North Ballaird Farm [NX 121 879] where sphalerite in carbonate veins locally elevates the Zn level in whole rock analyses to 3560 ppm. Arsenopyrite has a similar mode of occurrence but is present in only trace amounts.

Chrome mineralisation

In the northern serpentinite belt serpentinised harzburgite and dunite may, in places, contain abundant 1–1.5 mm-sized spinel grains. At one locality [NX 1386 9152] these are concentrated into a zone 4–5 m wide exposed in the raised sea cliff north of Pinbain Bridge (Figure 10). Variation in chrome spinel concentration from about 30 to 90 per cent causes subvertical layering approximately parallel to the nearby, faulted margin of the serpentinite belt close to which the chromitite is extensively brecciated. In thin section the spinel has a deep chestnut brown colour in transmitted light with reflected light showing that individual grains are fractured and embayed with thin, incomplete rims of magnetite. The texture of the rock, in which the accompanying bastite pseudomorphs after enstatite sometimes poikilitically enclose serpentinised olivine crystals, attest to an origin for the rock by crystal settling and accumulation. Electron microprobe analysis (Column 1, Table 2) shows the spinel to be true chromite containing about 51 per cent Cr_2O_3.

The spectacular nodular chrome spinel lithology (Figure 8) cropping out at the head of Poundland Burn in the southern serpentinite belt [NX 1698 8820] has been described by Teall (in Peach and Horne, 1899) and more recently by Jones (1977). Spinel grain size varies across the exposure (about 10 m) from 1 mm to 10 mm although the larger nodules may have formed by the coelescence of several smaller grains. The fractured and embayed spinels are contained in a serpentinised dunite host and at their most abundant form about 70 per cent of the rock. Electron microprobe analysis (Column 2, Table 2) shows the spinel to be deficient in Cr_2O_3 (about 38 per cent) compared to true chromites and

Table 2 Selected electron microprobe analyses of chrome spinels from the Ballantrae Complex

	Pinbain Bridge	Poundland Burn*	Poundland Burn+	Locality 10=	Locality 11=
SiO_2	0.16	—	—	0.33	0.69
TiO_2	0.38	0.23	0.43	—	0.30
Al_2O_3	20.14	29.91	26.87	17.60	19.30
Cr_2O_3	51.09	37.87	34.26	51.60	44.70
Fe_2O_3	0.20	3.15	10.16	3.01	—
FeO	16.30	11.22	14.90	14.10	25.40
MnO	—	0.18	0.58	—	0.46
MgO	12.83	16.50	13.88	14.00	11.60
CaO	0.29	—	0.29	—	0.20
V_2O_3	—	—	0.20	0.32	—
Total	101.39	99.06	101.57	100.96	102.65

* Spinel from nodular lithology
\+ Accessory spinel from dunite
= See Figure 10

so the term 'picotite', applied by Teall may be more appropriate. He described the spinel-rich zones as veins but Jones (1977), as discussed earlier, considered this lithology to be part of an ophiolitic cumulate sequence, an interpretation with which the present authors concur. The isolation of the small exposure of this unique lithology is probably the result of tectonic disruption of the original cumulate sequence in such a way that the spinel-rich rock is in effect a large, undeformed block within sheared serpentinite.

There are no other exposed lithologies in the Ballantrae ultramafic suite where chrome spinels are more than accessory minerals. These probably show a range of compositions across the 'picotite'-chromite spectrum, as illustrated by the electron microprobe analyses of accessory chrome spinels shown in columns 3 to 5 of Table 2, and by analyses of spinels panned as heavy mineral concentrates from stream sediments (Stone and others, 1986). Generally both the Cr_2O_3 content of the spinels and the Cr:Fe ratios are highest in the concentrates derived from the northern serpentinite outcrop, with a median Cr_2O_3 content for the concentrates of 5.86%, $(100 \times Cr) : Fe = 56$, and indicate that metallurgical grade chromite may be widely present there. This is confirmed by microprobe analyses of individual panned spinels which show Cr_2O_3 content ranging up to 63 wt%. By contrast spinel concentrates derived from the southern serpentinite outcrop have a median Cr_2O_3 content of only 1.88% and $(100 \times Cr) : Fe = 20$.

Petrogenetic implications of chrome mineralisation

The molecular proportion ratio Cr/Cr + Al has been used by Dick and Bullen (1984) to discriminate between peridotites generated in abyssal and sub-arc environments. Using their criteria very little of the ultramafic rock at Ballantrae is unequivocally of oceanic crustal origin, that is, generated beneath a spreading ridge. The spinels from the northern belt generally support an origin in a sub-arc setting whilst, in the southern belt, only the nodular picotites from the Poundland area [NX 1698 8820] and spinels in the same general vicinity, that is that part of the outcrop believed to contain the transition zone of an ophiolite sequence, have completely abyssal characteristics. Elsewhere the Cr:Al ratios support a sub-arc origin and suggest that the southern serpentinite outcrop is polygenetic. This is in accord with the possibility (discussed earlier in this chapter in the section on basic igneous xenoliths in the southern serpentinite) that the transition zone ultramafics of an ophiolite complex, at present exposed in the Poundland area, were intruded by peridotitic crystal-rich 'magma', the combined ultramafic mass now forming the southern serpentinite outcrop of the Ballantrae Complex. However, a trend of increasing Cr content and Cr:Fe ratio with greater depth in the ultramafic sequence is not uncommon in ophiolite complexes. An alternative view of the serpentinite bodies at Ballantrae would therefore regard the northern belt as originating at greater depth that the southern belt; both within the same ophiolite assemblage.

3 Volcano-sedimentary sequences of Arenig age within the Ballantrae Complex

Basaltic pillow lavas, associated breccias and various interbedded sedimentary rocks form about half of the area of outcrop of the Ballantrae Complex. Peach and Horne (1899) regarded all of the volcanic rocks as part of a single, middle Arenig succession dated by a sparse graptolite fauna collected at a few localities. The widespread agglomerates they described have now been generally recognised as essentially sedimentary breccias. Assumptions of overall stratigraphic unity continued through later work and was formalised by Bailey and McCallien (1957) who divided the volcanic sequence into a lower, Knockdolian Spilitic Group and an upper, Downan Point Spilitic Group separated by serpentinite which was thought by them to have originated as a submarine ultramafic lava. More recently Lewis (1975), whilst not accepting Bailey and McCallien's stratigraphy, re-emphasised the unity of the volcanic succession and described a single sequence at least 5.2 km thick. However, the use of trace element ratios in basalts to discriminate between different tectonic settings for their original eruption (Pearce and Cann, 1973) led Wilkinson and Cann (1974) to propose that the basaltic rocks of the Ballantrae Complex were derived from disparate sources, were juxtaposed tectonically, and could not originate within a single stratigraphic sequence. Several studies of this aspect have since been made and are discussed as appropriate below. Further work on the graptolite faunas (Stone and Rushton, 1983), which are best related to the Australasian Arenig biostratigraphy (Table 3), showed significant age differences between different volcanic outcrops and demonstrated structural imbrication within some parts of the succession. Clearly the concept of a single, stratigraphically well ordered, volcano-sedimentary sequence is no longer tenable, and Bailey and McCallien's (1957) scheme cannot be upheld, but the precise geological relationships throughout the outcrop are still far from clear.

For stratigraphical convenience in this account, all of the Arenig volcano-sedimentary sequences are brought together in the *Balcreuchan Group.* The name is taken from that part of the Bennane Head area [NX 095 875] where the volcano-sedimentary rocks are well exposed and the biostratigraphy and structural imbrication are best understood. Incidentally, the location was also the home of Sawney Bean, the notorious 16th century Scottish cannibal, who showed considerable geological acumen by choosing to live on ophiolitic rocks, an 'indigestible' section of oceanic crust. The component parts of the Balcreuchan Group are described and discussed by area of outcrop, sequentially from north to south (Figure 11). Where lithologies are sufficiently distinctive formal formations are proposed and are delineated on the accompanying map. The most southerly outcrop of volcanic lavas, here defined as the Downan Point Lava Formation, has an ambiguous situation and may be a part of the Southern Uplands imbricate thrust succession (Legget and others, 1979) rather than a member of the Ballantrae Complex (*s.s.*). The justification for and significance of this association is discussed fully below in Chapter 5.

The Balcreuchan Group: northern outcrop, Pinbain

The Pinbain block is that area of moderate relief dominated by Pinbain Hill [NX 1495 9205; Figure 11] and extending between Pinbain Bridge [NX 1375 9142], Kennedy's Pass [NX 1484 9315] and Grey Hill [NX 1646 9370]. It is formed of steeply dipping, massive and pillowed basic lavas and interbedded volcaniclastic sedimentary rock, all with a predominant ENE–WSW strike. Both dip and facing direction are consistently to the north-west except on Kilranny Hill [NX 1568 9248] where the local sequence is inverted and dips steeply to the south-east. Exposure is excellent in a coastal strip some 150–200 m wide but is poor inland, where detailed field relationships can often only be inferred. The Pinbain block is fault-bounded on all sides, bringing the volcanic sequence into contact with mafic and ultramafic rocks of the northern serpentinite belt to the south-east, and

Table 3 Biostratigraphical subdivision of the Arenig Series (correlation approximate, from Cooper and Fortey, 1982, fig.2)

Australasia		North America	England and Wales (based principally on the Lake District sequence)
Yapeenian	Ya 1–2	8 *Isograptus*	*hirundo*
Castlemainian	Ca 2–3		
	Ca 1	7 *bifidus*	*gibberulus*
Chewtonian	Ch 1–2	6 *protobifidus*	*nitidus*
Bendigonian	Be 3–4	5 *fruticosus* – 3 stipe	*deflexus*
	Be 1–2	4 *fruticosus* – 4 stipe	
Lancefieldian	La 3	3 *approximatus*	(not recognised)

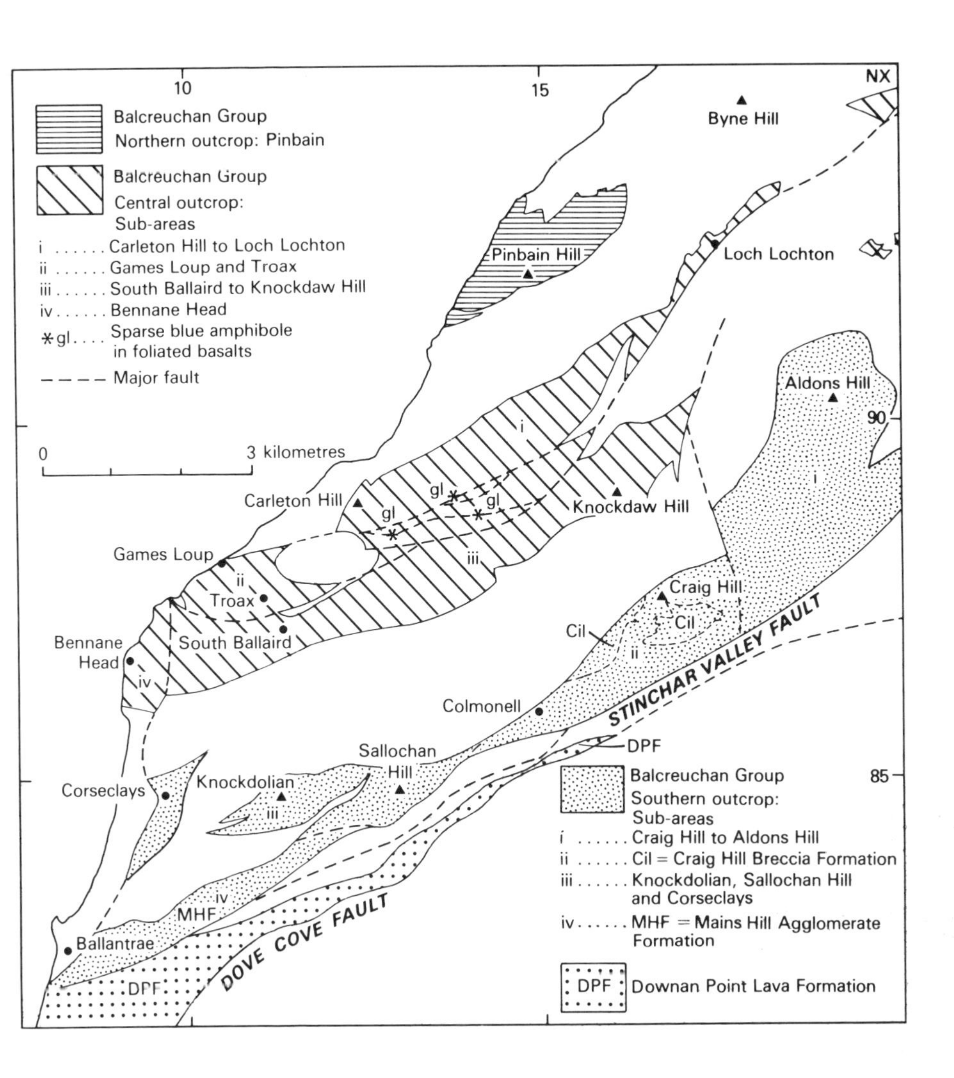

Figure 11 Outcrop distribution of the Balcreuchan Group within the Ballantrae Complex and of the neighbouring Downan Point Lava Formation

Llandeilian conglomerates of the Barr Group succession to the north-west.

Only two fossiliferous localities are known: both in the vicinity of Pinbain Bridge. Around Pinbain Bridge itself a structurally isolated slice of lavas and conglomerate includes a black mudstone mélange containing blocks of unsheared black mudstone and shale. These blocks have yielded only fragmentary fossils that are not diagnostic of age. They include Peach and Horne's (1899, p.442) record of *Climacograptus confertus*, which in fact refers to an undeterminable fossil (Rushton and others, 1986). The roadside exposure 180 m north of Pinbain Bridge exposes interbedded dark mudstone and sandstone (Peach and Horne, 1899, p.444) from which several faunas have been recovered, including the graptolite *Tetragraptus decipiens* T. S. Hall. These indicate an early Arenig age, probably equivalent to the Bendigonian (Be 1 or Be 2) of the Australasian sequence (Rushton and others, 1986) for the lower part of the Pinbain sequence (Figure 12). Despite examination of several other potentially fossiliferous sites, including interpillow limestone and limestone blocks from the muddy mélange deposits, no other exposure of the Pinbain sequence has thus far yielded diagnostic fossils.

The lavas are variably porphyritic and aphyric, although the aphyric lavas commonly contain tiny scattered micro-phenocrysts. Plagioclase is ubiquitous, and there are also rare augite and olivine phenocrysts set in a matrix of plagio-clase, augite, iron oxide, and, rarely, olivine. Alteration pro-cesses have widely converted this primary mineralogy to a secondary assemblage of albite, chlorite, sphene, epidote, prehnite and pumpellyite (prehnite-pumpellyite facies) in the south-eastern half of the area, and to less completely

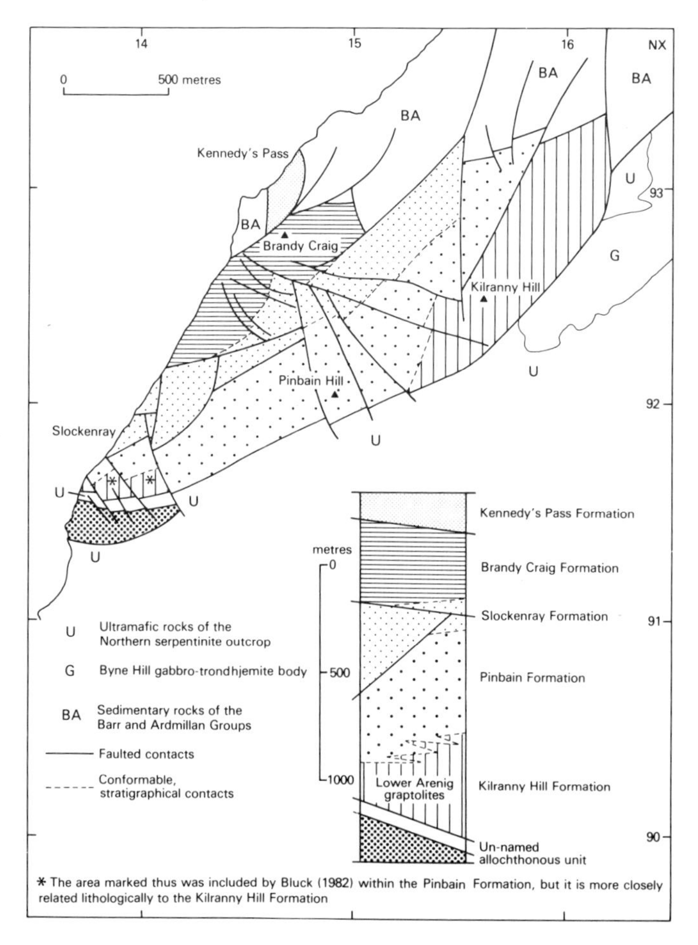

Figure 12 Lithostratigraphy of the northern outcrop of the Balcreuchan Group. Nomenclature after Bluck (1982). See Table 4 for further details

Table 4 Lithostratigraphy of the northern (Pinbain) outcrop of the Balcreuchan Group (cf. Bluck, 1982)

Formation	Approximate maximum thickness (m)	Lithological characteristics
Kennedy's Pass	115	Non-stratified, monomict, brown and green, aphyric lava-breccia with occasional isolated pillows and intact lava-masses. Some gravelly and strikingly porphyritic hyaloclastite similar to Slockenray Formation examples.
Brandy Craig	285	Pillowed and massive, porphyritic and aphyric lavas, often intimately intermixed and with local reddened horizons suggestive of subaerial weathering. Minor thin discontinuous beds and lenses of polymict lava breccia, lithic arenite, laminated siltstone, tuff and porphyritic and aphyric hyaloclastite.
Slockenray	285	Massive, porphyritic and aphyric hyaloclastites and porphyritic and aphyric massive lavas (intimately intermixed in part) made distinctive by the presence of large (to 2 cm) bladed plagioclase phenocrysts. Some thin-bedded lithic arenite and granule- to pebble-conglomerate.
Pinbain	475	Thick, massive and pillowed, porphyritic and aphyric lavas, brecciated in part and of multiple-flow origin. Rare discontinuous beds and pockets of fine lithic arenite, siltstone, mudstone and polymict lava breccia, some with dispersed intact lava pillows.
Kilranny Hill	300	Massive and faintly-bedded, fine to very coarse crystal-lithic arenites, laminated silty and cherty mudstones, well bedded arenites, and occasional polymict pebble- to boulder-conglomerate. Locally interdigitates with thin aphyric and porphyritic lavas of the Pinbain Formation.
Un-named allochthonous unit	180	Disorganised, polymict breccio-conglomerates and thin black mudstone mélange deposits with exceptionally wide range of ophiolite-derived clasts (cf. Bailey and McCallien, 1957; Church and Gayer, 1973; Bluck 1978). Small outcrops of monomict gabbro-breccia, aphyric pillow-lava and pillow-disintegration breccia have uncertain relationships.

recrystallised assemblages comprising analcime, zeolite, smectite and albite (zeolite facies) to the north-west (Figure 13). However numerous exceptions to the broad distribution pattern are known (Smellie, 1984a; Oliver and others, 1984).

Unlike other parts of the volcano-sedimentary outcrop elsewhere in the Ballantrae Complex, the generally adequate exposure and the presence of distinctive, locally mappable rock types within the Pinbain block allow a continuous lithological pattern to be distinguished (Bluck, 1982, fig.1). Five named units of formation status and a sixth un-named unit within a fault-bounded allochthonous block have been identified (Figure 12; Table 4). Along the coast, strike faults appear to separate most of the units distinguished, but conformable, depositional contacts are known inland. Interdigitation of sediments of the lowermost, Kilranny Hill Formation with lavas of the overlying Pinbain Formation can be inferred from relationships exposed in streams draining the north-eastern flanks of Kilranny Hill. The un-named unit, entirely enclosed by serpentinite at the south-western margin of the Pinbain outcrop, contains the well known Pinbain beach section of ophiolite-derived and blueschist-bearing (Ed 6129–6131) sedimentary mélange deposits (Figure 14). These have been described in detail by several authors (Peach and Horne, 1899; Bailey and McCallien, 1957; Bluck, 1978) who, in addition to the blueschist, report clasts of epidote schist, gabbro, basalt, dolerite, pyrite, pyroxenite, amphibolite, limestone, chert, greywacke and shale. Despite this attention the facing direction of the subvertical ESE-striking bedding remains uncertain. The marked contrasts in lithology, provenance and bedding orientation relative to the adjacent Pinbain sequence suggests that the un-named unit is allochthonous. However, it can be closely matched with sedimentary lithologies forming the north-eastern prolongation of the Carleton Hill to Loch Lochton sector (see the following section), from which it may have been derived by strike-faulting.

Depth of water during deposition of the un-named allochthonous unit is unknown. The presence of black, graptolitic mudstones suggests quiet, deep-water conditions, but the rounding displayed by many clasts within the interbedded conglomerates requires either derivation from a conglomeratic provenance or that the clasts achieved their rounding in a higher-energy, probably beach environment prior to resedimentation in deeper water. Conversely, evidence for shallow-water deposition is widespread in the upper part of the main Pinbain sequence, particularly within the Slockenray and Brandy Craig formations, where intertidal hyaloclastite delta deposits, possible subaerial lava weathering, and subaerial ash-fall constituents in sedimen-

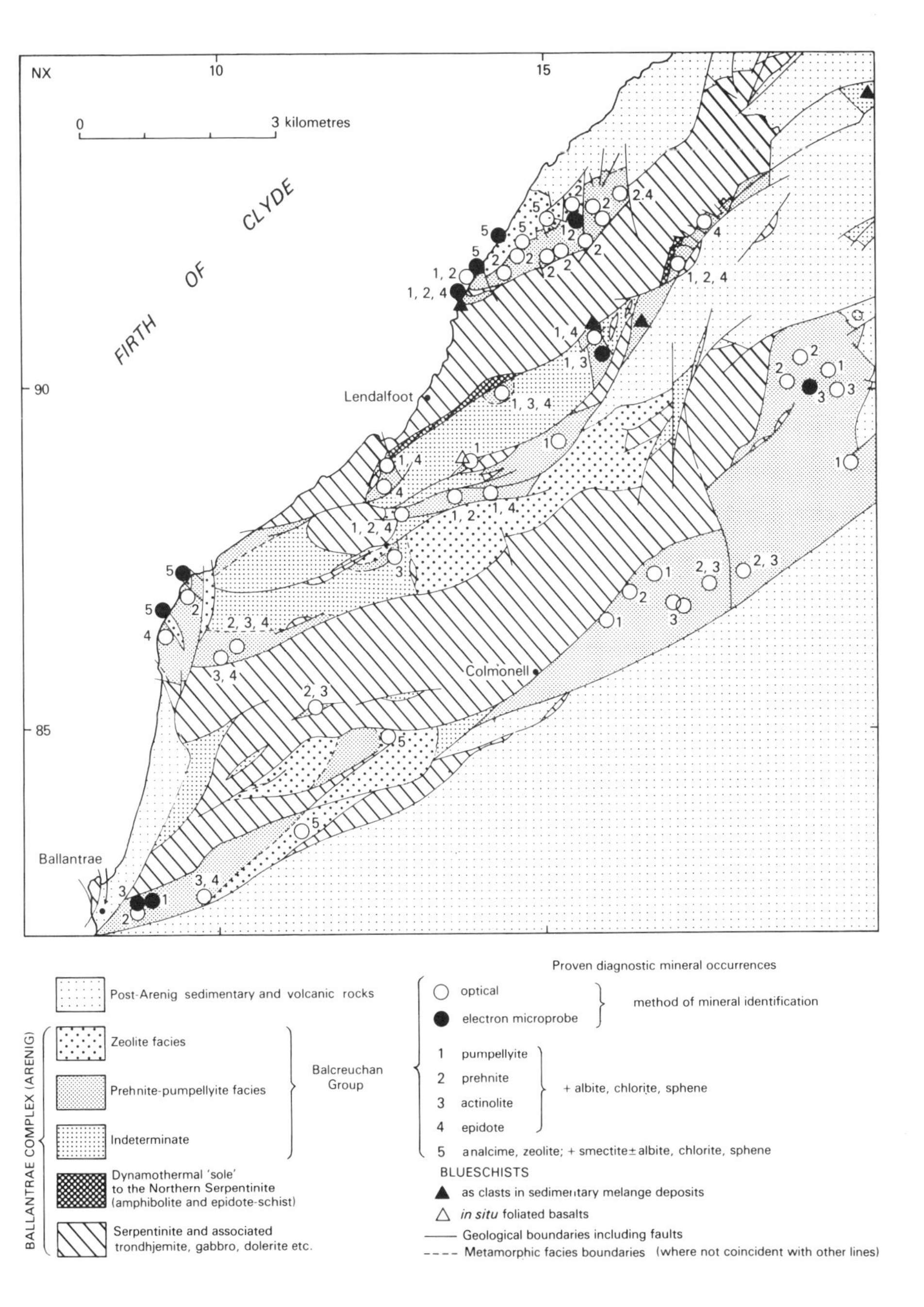

Figure 13 Metamorphic facies distribution within the outcrop of the Balcreuchan Group

tary rocks have all been identified (Bluck, 1982; Smellie, 1984b).

Although many of the lavas appear, from the map, to attain considerable individual thickness (up to 250 m), the outcrops invariably contain thin, discontinuous bands and lenses of volcaniclastic sedimentary rock attesting to a multiple-flow origin. Moreover, many lavas, especially within the Slockenray and Brandy Craig formations, are characterised by the intimate intermixing of porphyritic and aphyric lavas (with either type dominant) due to the simultaneous extrusion of the two types. Based principally on major oxide chemistry and presumed stratigraphical relationships, Lewis (1975) and Jelinek and others (1982) assigned the Pinbain lavas to eruption in an island arc, or more general arc/marginal basin setting, respectively. How-

Figure 14
Variation in clast lithology in the Pinbain mélange deposit. The clasts are enclosed in a pervasively foliated, mudstone matrix. D 3525

ever, these interpretations have been criticised by Jones (1977) and Thirlwall and Bluck (1984) who showed the lavas to be typical ocean tholeiites, with high Fe/Mg, V and Sc, low Ni and Cr, and Sr-Nd isotopes characteristic of oceanic island ('hot-spot' or within-plate) basalts such as those from Hawaii. Lava clasts from within the mélange units on Pinbain beach have a chemistry transitional between alkalic and tholeiitic basalt, and some closely resemble modern alkali basalts in their enriched stable-element pattern; generation within an oceanic island environment is also envisaged for these rocks (Wilkinson and Cann, 1974; Thirlwall and Bluck, 1984).

The Balcreuchan Group: central outcrop

Between the two main serpentinite belts the central outcrop of the Balcreuchan Group is divided by major structures into four discrete subareas (Figure 11). A major shear zone, coincident with the south-western end of the central zone of ultramafic inliers (Figure 2) is the dominant feature, dividing the Carleton Hill/Loch Lochton and Games Loup/Troax sectors on the north-west from the South Ballaird/Knockdaw Hill sector to the south-east. Within the shear zone foliated basalts locally contain sparse blue amphibole (gl on Figure 11). That part of the Balcreuchan Group outcrop to the north-west of the shear zone is further divided at its south-west end by the thrust slice of serpentinite at Balsalloch (G on Figure 2). Structurally above the thrust serpentinite the Games Loup/Troax outcrop is thus distinct from the Carleton Hill/Loch Lochton sector farther north-east which structurally underlies the serpentinite. At the opposite, south-west, end of the Balcreuchan Group's central outcrop, a major north–south fault isolates the Bennane Head subarea from the South Ballaird/Knockdaw Hill sector. The importance of all of these boundary structures is emphasised by the differences in the Balcreuchan Group volcano-sedimentary sequences evident between adjacent subareas. These will be described below in order from north to south.

Carleton Hill to Loch Lochton

The north-western faces of Carleton and Balsalloch hills are formed by part of the metamorphic aureole at the base of the northern serpentinite belt. Metasedimentary rocks within the aureole merge south-eastward with the unaffected part of the volcano-sedimentary sequence which is well exposed on the summits and southern flanks of the hills. The lavas and interbedded sediments maintain a fairly uniform attitude with NE–SW strike, steep dip towards the north-west and a consistent north-west facing direction. The lavas are chiefly aphyric, consisting essentially of a fine-grained mesh of albitised plagioclase, augite and chlorite, but small plagioclase phenocrysts are by no means uncommon to give porphyritic textures locally. The same lithological range is seen in the abundant interbedded pillow disintegration breccias (the macadam breccias of Lewis and Bloxam, 1980) which contain angular to subrounded basalt clasts with virtually no enclosing matrix. Small, NW-trending dykes (generally less than 30 cm wide) cut and are chilled against the lavas and breccias with which, however, they are petrologically identical. The dykes are interpreted as feeders to the upper part of the lava sequence and therefore were approximately coeval with the extrusive basalts. The inter-pillow sediment in this area, apart from the breccias, includes black siliceous mudstone, tuff, chert and thin shaly limestone, all of which have proved unfossiliferous. Towards the north-west margin of the outcrop the proportion of this material in the sequence increases to the extent that there are lithologically mixed sedimentary horizons at least 20 m thick on both the north-west and south-east faces of Carleton and Balsalloch hills. A similar thickness of red mudstone crops out farther east in the banks of the Water of Lendal to the south-west of Cundry Mains Farm [NX 145 899]. There is abundant evidence locally for slumping within these sedimentary sequences.

A steep north-westward dip and NE–SW strike continue thoughout the north-easterly prolongation of the volcano-sedimentary outcrop, despite frequent strike- and cross-faulting; the facing direction, however, is unknown. Massive aphyric or sparsely porphyritic lavas are predominant but are pillowed in parts, such as the hillside 450 m NNE of Currarie [NX 1622 9094]. Scrappy exposures of feldsparphyric lava occur in the banks of the Water of Lendal upstream from Straid Bridge [NX 1390 9005] and in hillslopes north-west and south-west of Knocklaugh [NX 1727 9180], but the only major exposure is restricted to a fault-bounded block at Pennyland Wood [NX 1570 9065]. The north-western margin of the volcano-sedimentary sequence, adjacent to the northern serpentinite belt, is formed principally by black and grey, sheared mudstone mélange. The mélange outcrop is characterised by hummocky, grass-covered hillsides dotted by protruding, more resistant blocks ('knockers') of aphyric lava. However, in the stream sections (especially Lochton Burn [NX 1713 9208] and streams around Currarie [NX 1586 9102 and NX 1632 9101]), a more varied clast assemblage is seen. Conspicuous white limestone blocks are common, accompanied by clasts of serpentinous gabbro, dolerite, serpentinite, coarse sandstone, epidote amphibolite, mica schist and blueschist. In two places, the deposit is a fine breccia formed exclusively of angular, close-packed fragments of blueschist (S 69963, 70408) or mica-schist [NX 1569 9101 and NX 1813 9309, respectively].

Lewis (1975) considered that the basaltic lavas of Carleton and Balsalloch hills showed moderate iron-enrichment and major element ratios typical of island-arc tholeiites. His information on the Pennyland Wood/Loch Lochton area was more scanty, however the lavas were said to be broadly comparable to other lavas of island-arc tholeiitic chemistry exposed elsewhere in the Complex. From a study of the Ti-Zr-Y ratios Jones (1977) preferred an origin for all of the lavas as oceanic crust generated at a spreading ridge. Subsequently Lewis and Bloxam (1980) reinforced the island-arc interpretation by comparing the high proportion of breccias in the sequence to the similar situation in modern island arcs: a contrast to the general paucity of breccia apparent in modern oceanic ridge environments. Much the same argument had been earlier used by Bluck (1978) and subsequently by Stone (1982), but it is by no means conclusive.

Throughout the south-western part of the Carleton Hill/Loch Lochton subarea a spatially irregular development of secondary prehnite, pumpellyite and epidote, together with more ubiquitous chlorite, suggests that much of the sequence has experienced burial metamorphism in the prehnite-pumpellyite facies. North-eastward the diagnostic mineral assemblage becomes less well defined, although the actinolite noted locally confirms the continuation of the same general metamorphic grade. There are however large areas devoid of relevant petrographical information and so the overall distribution of metamorphic facies remains uncertain (Figure 13; cf. Oliver and others, 1984; Smellie, 1984a).

As the metamorphic grade of the Carleton and Balsalloch hills basalts becomes less well defined north-eastwards it is further complicated between Carleton Mains Farm [NX 1375 8960] and Knockormal Farm [NX 1373 8864] by sheared basalts containing sparse blue amphibole (Figure 10) adjacent to the Knockormal mafic-ultramafic body (see the description of the central zone in Chapter 2). The basalts are, however, partially enclosed as a tectonic inclusion within serpentinised ultramafic rock on the northern side of the fault zone and so many possibly be unrelated to the adjacent volcano-sedimentary sequence. On the evidence of their trace-element ratios Wilkinson and Cann (1974) described the blue-amphibole-bearing schistose basalts as ocean-floor tholeiites. The blue amphibole has been described by Bloxam and Allen (1960) as both glaucophane and crossite, and it occurs with abundant green amphibole in schists interbanded with epidote amphibolites lacking the blue varieties. Typically (S 67601) the larger crystals of green, hornblendic amphibole in glaucophane- and crossite-bearing lithologies are fringed by the blue amphiboles. In this respect the Knockormal and Carleton Mains 'blueschists' are distinct from the blueschist clasts in the Pinbain and other mélange deposits (some of which are in addition garnetiferous) which contain very little green amphibole and no fringing relationships. Thus the Pinbain and other deposits do not appear to have been derived from the Knockormal/Carleton Mains rocks. The status and significance of the Knockormal/Carleton Mains 'blueschist' lithologies has been disputed (J. E. Dixon quoted in Smellie and Stone, 1984) but they do appear to be in situ. Their origin was most likely within a deep shear zone adjacent to the central zone of ultramafic 'inliers' within the Ballantrae Complex. Their presence emphasises the importance of this structural zone as one of the fundamental features of the Complex.

Games Loup and Troax areas

At the south-west extremity of the outcrop on land of the northern serpentinite belt, the coastal exposure at Games Loup (Gemmells Loup in some older literature) [NX 103 880] the ultramafic rock is faulted against a sequence of well pillowed basaltic lavas. These are exceptionally well exposed south-westward along the coast for about 0.5 km and are then juxtaposed with a markedly different volcano-sedimentary sequence (the Bennane Head area discussed below) across the major fault zone at Balcreuchan Port [NX 099 877]. The Games Loup lavas strike approximately NE–SW, parallel to the coast, so that the same stratigraphic level is exposed at all points of the coastal section. The attitude of the lavas is invariably steep and younging evidence consistently indicates that the sequence faces north-west. Thus the rocks exposed on the coast are, in local terms, the uppermost part of the volcano-sedimentary sequence. No evidence was found during the recent survey work to support the proposal by Mendum (1968) that within the Games Loup section two lava units with opposing facing directions were separated by an angular unconformity.

The lavas exposed in the Games Loup coastal section typically contain small phenocrysts of augite and more rarely of albitic plagioclase in a fine-grained groundmass of albite, clinopyroxene and chlorite. Sparse vesicles are commonly filled by calcite, chlorite or bowlingite, the bowlingite also forming pseudomorphs, in association with serpentine, after olivine phenocrysts (S 69609). Spilitisation has undoubtedly caused the albitisation of originally more calcic plagioclase but the presence of an additional secondary mineral assemblage, locally including sphene, epidote and prehnite, suggests burial metamorphism under conditions approximating to the prehnite-pumpellyite facies (Smellie, 1984a). However, the augite phenocrysts are generally fresh and unaltered and have given a Sm-Nd age of 476 ± 14 Ma (Thirlwall and Bluck, 1984).

North – south jointing and minor faulting both affect the coastal section but there is no evidence for a major structural discontinuity between the Games Loup lavas and the poorly exposed and weathered volcano-sedimentary succession farther inland. The attitude and facing direction of the strata there are the same as in the coastal section suggesting that the rocks underlie those exposed on the coast. The Games Loup lavas contain virtually no interpillow sediment and are generally free of brecciation. By contrast the inland, stratigraphically lower succession contains a high proportion of basalt breccia interbedded with the lavas; a proportion which apparently increases down sequence towards the south-east. There is also an along-strike increase in clastic proportion north-eastward, so that the eastern side of Troax Hill [NX 114 878] is entirely formed of oligomict basalt breccias with small (less than 10 cm in diameter) subrounded and subangular clasts; some also contain rare clasts of chert. The basalts and the basalt clasts are all aphyric, composed essentially of abite, clinopyroxene and chlorite with accessory secondary epidote and sphene. Neither clinopyroxene phenocrysts nor secondary prehnite were observed, in marked contrast to the stratigraphically higher coastal sequence.

The breccias are reddened along the eastern margin of their outcrop close to the thrust contact with the Balsalloch serpentinite body. In this general zone, but adjacent to the faulted southern margin of the Games Loup/Troax succession, a sequence of clastic sediments and breccias overlying silicified serpentinite, nowhere exposed, has been described from a borehole near North Ballaird Farm [NX 1210 8777] by Stone and Strachan (1981). The core recovered contained an abundant isograptid fauna (Figure 15) of Castlemainian or Yapeenian (Upper Arenig) age (Stone and Rushton, 1983). This age compares more or less favourably with the Sm-Nd age of 476 ± 14 Ma obtained by Thirlwall and Bluck (1984) from augite phenocrysts in basalts of the coastal section, depending upon the time scale used (e.g. Harland and others 1982, McKerrow and others 1985). The precise status and relationships of the borehole sequence remain enigmatic.

Geochemical work in this area has been restricted to the well exposed and less pervasively weathered coastal sequence of clinopyroxene-phyric basalts. From a consideration of the Ti, Zr, Y and Nb ratios Wilkinson and Cann (1974) described the lavas as typical low-K tholeiites of island-arc origin. Lewis (1975) studied the major element characteristics and described the lavas as basaltic andesites but agreed with the likely island-arc origin. This was further reinforced by a subsequent study (Lewis and Bloxam, 1977) of the rare-earth elements, although the final choice between an island

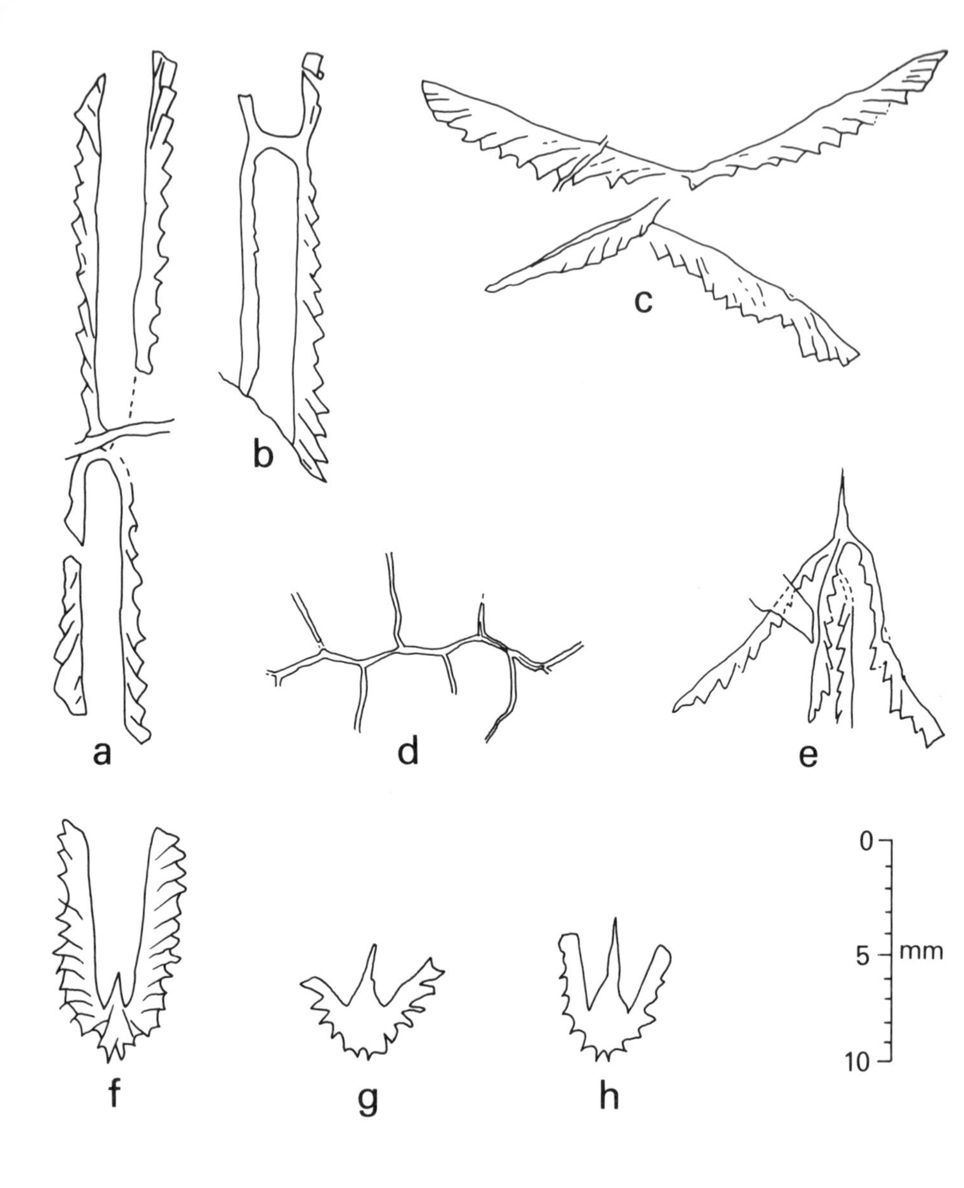

a, b ..... *Tetragraptus (Paratetragraptus) approximatus* Nicholson. GSE 13824 and 13825. Lower Arenig. Bennane Head, west of Balcreuchan Burn.

c ..... *Tetragraptus (T.) reclinatus reclinatus* Elles and Wood. GSE 13830. Middle Arenig. Bennane Head, Balcreuchan Burn-foot.

d ..... *Sigmagraptus praecursor* Ruedemann. GSE 13860. Lower—Middle Arenig. Bennane Head.

e ..... *Tetragraptus (Pendeograptus) fruticosus* (J. Hall). GSE 13828. Lower Arenig. Bennane Head, west of Balcreuchan Port.

f ..... *Isograptus caduceus* (Salter) s.l. GSE 13727. Upper Arenig. Borehole at North Ballaird.

g, h ..... *Pseudisograptus dumosus* (Harris). GSE 13724 and 13725a. Upper Arenig. Borehole at North Ballaird.

Figure 15 Graptolites from the Balcreuchan Group. Numbers prefixed GSE refer to specimens housed in the collection of the British Geological Survey, Edinburgh

arc and a spreading ridge source was strongly influenced by erroneous stratigraphical considerations. A different origin was proposed by Jones (1977) from a study of the Ti-Zr-Y relationships which he considered more typical of within-plate 'Hawiian-type' basalts. More recently the island-arc origin has received further support from Thirlwall and Bluck (1984), who liken the chemistry of the Games Loup lavas to that of modern, primitive intraoceanic island-arc tholeiites: high in MgO, Ni and Cr but depleted in some rare-earth elements, Zr, Nb, Ti and P.

South Ballaird to Knockdaw Hill

The steep dip and generalised NE–SW strike of the lavas to the north is maintained throughout the 7 km length of the Balcreuchan Group outcrop to the immediate south-east of the central zone of serpentinite bodies (Figure 11). However, there is local reversal of the dominant north-west facing direction, principally in the vicinity of the new South Ballaird Farm [NX 114 871]. The level of exposure is variable, poor over large areas and excellent on some of the higher ground, where the rock seen is dominantly a uniformly aphyric, usually vesicular, dark grey-green and well pillowed lava. In thin section the lava is seen to consist of a randomly orientated mass of sericitised plagioclase, chlorite or smectite, with some augite and occasional microphenocrysts of plagioclase. Pyrite and ilmenite are common accessories and some of the chlorite forms pseudomorphs possibly after orthopyroxene. Trachytic texture is developed only locally. In the western part of the area there is some alteration of the augite to actinolite with secondary epidote and sphene also commonly present. However, eastward in the vicinity of Knockdaw Hill [NX 163 889] the burial metamorphic grade seems slightly lower, with no actinolite present, smectite very much more abundant than chlorite and only limited albitisation of the plagioclase. This is reflected in a relative eastward increase in the $CaO:Na_2O$ ratio in whole-rock analyses.

Interpillow sediment is common, particularly in the western part of the outcrop, and may be either chert, mudstone, tuffaceous sandstone or limestone. In places these lithologies form discrete horizons, a notable example being a limestone several metres thick cropping out on the north side of Knockormal Hill [NX 135 886], but no fossils were found in any of these sedimentary units. Pillow breccias are very common, especially in the Knockormal Hill/Moak Hill sector, and consist entirely of lava debris; angular to subrounded clasts packed together with virtually no matrix, forming typical 'macadam-breccias' as described by Lewis and Bloxam (1980). Entire lava pillows may be enclosed within these breccias; on the western side of Lochton Hill a transition is seen from breccia containing rare whole pillows to in-situ but entirely brecciated well pillowed lavas.

An unusually thick sequence of tuff and black siliceous mudstone, perhaps as much as 30 m, crops out in the South Ballaird/Lochton Hill area, where it is locally associated with conglomerate. The tuffs consist of a variety of fine-grained lava debris, including much glassy material, and locally contain celadonite as a vesicle filling (S 67518). In contrast to the more widespread oligomict pillow disintegration breccias the conglomerates contain a variety of clast types in a heterogeneous, sandy matrix. Balsillie (1937) described the subrounded clasts as chiefly of spilitic basalt accompanied by some gabbro and dolerite, granulite, amphibolite and serpentinite. The presence of all of these lithologies was confirmed during the recent survey, the pyroxene-hornblende granulites described by Balsillie probably being the same as the beerbachite contained as xenoliths within the southern serpentinite body. The association of such a conglomerate with abyssal black shales is likely to be the result of slumping and the conglomerate has been described as an olistostrome by both Church and Gayer (1973) and Bluck (1978), derived by the erosion of an ophiolite complex. The range of clast types suggests that the supposed ophiolitic source was structurally complex or deeply eroded. If this source is now represented by the other components of the Ballantrae Complex then the conglomerates, and by implication the associated shales, tuffs and lavas, must be amongst the youngest rocks now exposed therein. Another thick succession of clastic rocks was encountered in a borehole sited approximately 800 m east of Balsalloch Farm (Stone, 1982), consisting of conglomerate (with subrounded pebbles of basalt, gabbro, greywacke and chert), tuffs rich in glass shards and interbedded, chaotically slumped siltstone and shale. Overall it seems probably that, in the Ballaird–Balsalloch area, a far higher proportion of the succession consists of clastic sedimentary rocks, many derived by mass flow, than is immediately apparent from the larger exposures. The lava sequence showing reversal of facing direction at the new South Ballaird Farm [NX 114 871] is in close proximity to the slump conglomerates and so may in fact be contained within a disturbed unit with no tectonic implications.

On chemical grounds Lewis (1975) and Lewis and Bloxam (1977) argued for a conformable sequence 1.7 km thick uncomplicated by folding or tectonic repetition. Lewis described a systematic 'stratigraphic' variation in a number of element abundances which he considered compatible with a typical tholeiitic Fe-enrichment differentiation trend. From their chemistry, and the lack of free olivine, the lavas were further defined as low-K quartz tholeiites. The element ratios are compatible with an origin at either an island-arc or an oceanic-spreading ridge, but in view of the apparently great thickness of the sequence Lewis and Bloxam favoured an island arc. In contrast Jones (1977) interpreted chemical data from the lavas as indicating that two distinct sequences were present divided by a line somewhere between Moak and Knockdaw hills. To the west of this line Jones considered the lavas to have the Ti-Zr-Y ratios of within plate, oceanic island basalts whereas to the east, the Knockdaw Hill lavas had affinities with MORB-type oceanic crust. The two suites could be separated by a continuation of one branch of the anastomosing Knockormal fault complex extended to merge with the contact fault separating serpentinite and pillow lava on the south side of Knockdaw Hill.

Within this same fault zone blue amphibole occurs in sheared basalts (S 67277) adjacent to the southern margin of the Knockormal mafic-ultramafic fault slice (Bloxam and Allen 1960). The blue amphibole is principally crossite and occurs as fringes around larger crystals of hornblende in a similar fashion to the better developed examples occurring to the north of the Knockormal ultramafic slice (described earlier in this chapter in the section on Carleton Hill and Loch Lochton). Pumpellyite has also been described from the Knockormal area as an accessory mineral in quartz veins and more rarely, together with epidote, in calcite veins (Bloxam, 1958). Both pumpellyite and epidote were confirmed as secondary minerals in the basalts during the recent survey and, in addition, in the Knockormal area were found to be associated with prehnite. A prehnite-pumpellyite burial metamorphic grade is therefore likely for the basalts at Knockormal but the situation is confused by an apparent westward gradation into a lower metamorphic grade (Figure 13) with smectite rather than chlorite present in the basalt matrix (Smellie, 1984a).

Bennane Head

The excellent coastal exposures between Balcreuchan Port [NX 099 877] in the north and the Permian outcrops south of Bennane Lea [NX 092 859] form the best known section through the Balcreuchan Group volcanic and associated sedimentary rocks (Figure 16). In the northern part of the section the steeply inclined beds strike approximately north–south and face consistently towards the west. However, south of Bennane Head the strike changes to an east–west trend with a younging of the sequence towards the south. Lewis (1975) deduced a conformable stratigraphic succession about 2.4 km thick, an estimate subsequently halved by Barrett and others (1981), but Stone and Rushton (1983) were able to demonstrate a contradiction between the younging evidence and the distribution of early and middle

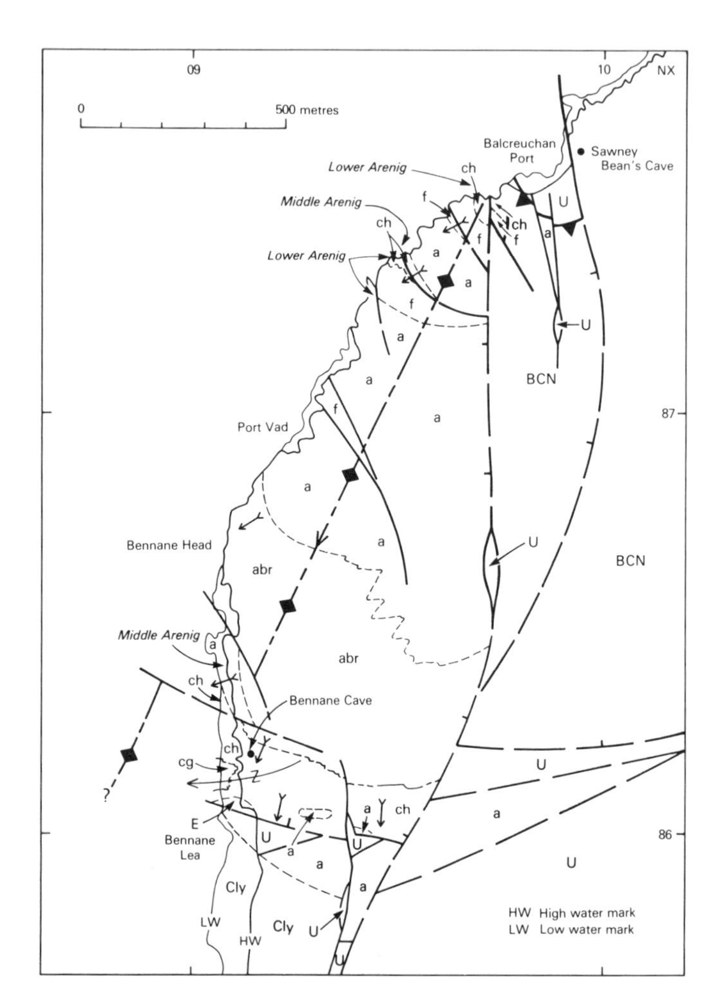

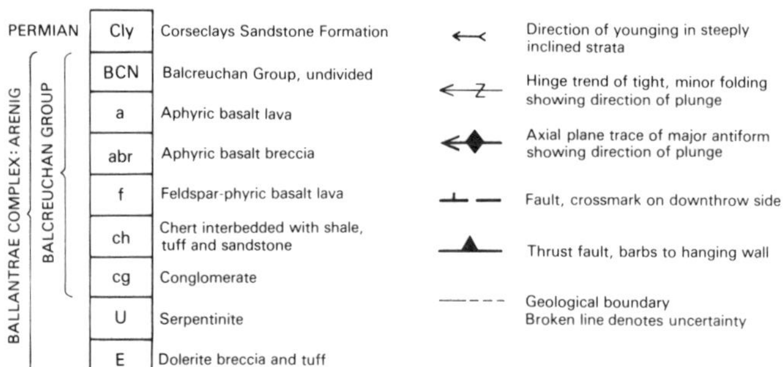

Figure 16 Outline geology of the Bennane Head area

Arenig graptolite faunas which can be most readily explained by the tectonic repetition of the succession.

The repeated sequence (summarised in Table 5) consists essentially of a basal unit of cherts and clastic sedimentary rocks at least 30 m thick (the base is never seen) overlain conformably by about 100 m of reddened, markedly feldsparphyric and well pillowed lavas with abundant thin interbedded chert horizons and red cherty interpillow material (Figure 17). The sedimentary rocks both within and underlying the lavas contain lowermost Arenig graptolites (Figure 15), including *Tetragraptus approximatus* and *T. fruticosus* (Stone and Rushton, 1983). *T. approximatus* alone represents a Lancefieldian 3 age (Table 3) whilst the two forms together are typical of the Bendigonian Stage. Up to 35 per cent albitised plagioclase phenocrysts are found in the feldsparphyric lavas contained in a hematite impregnated matrix of albite, chlorite and augite. An abrupt change in lava type to aphyric basalt occurs above the feldsparphyric lavas but despite some bedding-parallel faulting the contact between

Table 5 A summarised stratigraphy for the Bennane Head sector of the Balcreuchan Group

Approximate thickness (m)	Lithology	Biostratigraphy (Table 3)	Location of exposure
Top not seen	*Fault*		
10 +	Conglomerate		Bennane Lea section
30	Bedded radiolarian chert		
5	Interbedded chert, sandstone and breccia		
100	Massive, aphyric basalt lavas with thin mudstone interbeds	$Be_3 - Ch_2$*	
200	Oligomict basalt breccia		Bennane Head
10	Sandstone, tuff and conglomerate	Chewtonian*	This section forms the repeated, imbricated succession between Port Vad and Balcreuchan Port
50	Oligomict basalt breccia		
100 – 200	Aphyric, basalt pillow lavas		
100 – 200	Feldsparphyric basalt pillow lavas		
30 +	Sandstone, chert and conglomerate	$La_3 - Be_1$	
Base not seen	*Fault*		

* There is a hint here of further structural complication in that the stratigraphically apparently higher of these two faunas may be older than the lower. The biostratigraphy is however not sufficiently precise to allow firm conclusions to be drawn.

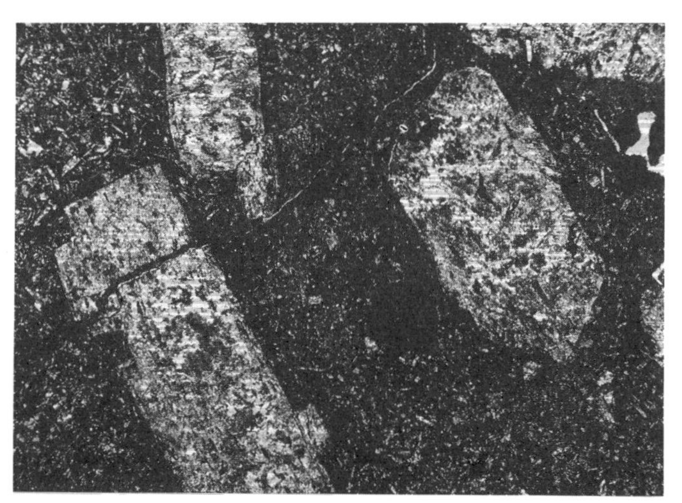

Figure 17 Large pillows of spilitic basalt protruding through a veneer of early Arenig chert (*top*); the basalt is markedly feldsparphyric (*bottom*): D 3587 and PMS 468, S 70430. Plane polarised light ×20 Bennane Head

the two types is exposed and perfectly conformable. The aphyric lavas (Figure 18) are unreddened, but are otherwise very similar in composition to the matrix of the feldsparphyric varieties. The aphyric lava unit is variously 100–200 m thick with an increasing brecciation trend upwards. Interbedded sediment is extremely rare, but conformably above the brecciated lavas occur several tens of metres of conglomerate, tuffaceous sandstone and shale, the uppermost horizons of which contain graptolites (Figure 15) such as *Tetragraptus reclinatus, T.* aff. *kindlei, Sigmagraptus* and *Didymograptus* cf. *protoindentus* of mid-Arenig age (Chewtonian Stage; Stone and Rushton, 1983). It may be significant that the older parts of this lava sequence contain rare

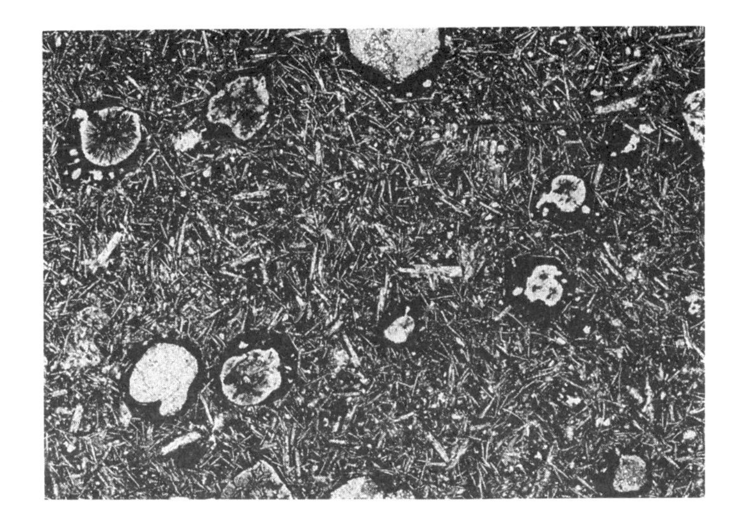

Figure 18 Aphyric basalt with amygdales of calcite and chlorite. Plane polarised light ×20 S 70428, PMS 464

prehnite, epidote and sphene as alteration products, suggesting burial metamorphism to the prehnite-pumpellyite facies, whereas the younger lavas contain zeolite and smectite, suggesting a lower metamorphic grade (Smellie, 1984a).

The succession described above, and summarised in Table 5, has been structurally imbricated along bedding-parallel thrusts so that, from a consideration of the biostratigraphy, at least three repetitions of all or part of the sequence crop out between Balcreuchan Port and Port Vad. Further evidence for imbrication is seen at Port Vad [NX 093 870], where aphyric lavas at zeolite grade are sandwiched between feldsparphyric lavas and more aphyric basalts, both in the prehnite-pumpellyite facies (Figure 13). The structural history of imbrication in this area is clearly complex and may possibly be polyphase.

South-west from Port Vad to the southern outcrop margin at Bennane Lea, direct evidence of imbrication is lacking and the succession continues to young uniformly westward with a predominant north–south strike and a steep dip. The proportion of oligomict basalt breccias increase rapidly and the Bennane Head/Bennane Hill area (Figure 16) is formed almost entirely of this lithology (Figure 19) in a succession several hundred metres thick with water-lain and fine-grained, basaltic sandstone interbeds. The clasts in the breccias are variably angular to rounded but the great majority are subangular in the size range 1–5 cm, with virtually no fine-grained matrix. Larger clasts tend to be more rounded. There are no clasts other than basaltic fragments, but there is considerable variation even within this limited range; highly vesiculated pumice contrasting with aphyric lava (Figure 19), and a clast colour variation from a reddish purple to

Figure 19 Oligomict basalt 'macadam' breccia. Note the angularity of the clasts, the presence of pumice fragments and the paucity of matrix. Bennane Head. MNS 3838

grey-green. Markedly feldsparphyric clasts are not present. Southward along the coast from Bennane Head the succession appears to continue conformably with the reappearance of aphyric basalt lava, here massive rather than pillowed, stratigraphically above the breccia. An overall upward stratigraphical progression is also supported by the presence of mid-Arenig graptolitic cherty mudstone interbedded with the lavas. The fauna (Figure 15) includes *Tetragraptus fruticosus, Sigmagraptus praecursor, Didymograptus extensus* and *D.* cf. *protomurchisoni,* and is considered to lie close to the Bendigonian – Chewtonian boundary (Stone and Rushton, 1983).

As the succession continues to young generally to the south-west, there is a marked change of strike south from Bennane Head so that in the Bennane Lea area [NX 092 859] bedding strikes almost east – west, maintaining its steep attitude (Figure 16). The sequence is now dominated by interbedded fine-grained basalt breccias and chert with an overall sense of younging to the south. Minor faulting confuses the relationship in the critical 'hinge' zone, but the varying bedding attitude of the volcano-sedimentary sequence in the whole Bennane Head area could be the result of its inclusion in a large, open antiform plunging moderately to the south-west (Figure 16; cf. Lewis, 1975). An alternative explanation favoured by Robertson (1982) involves an angular unconformity. In the southern limb of the fold (or above the unconformity) an unusual chert-conglomerate sequence is exposed on the shoreline in the vicinity of Bennane Cave [NX 092 863]. The bedded chert unit, several tens of metres thick with rare interbeds of basaltic sandstone, becomes progressively more disturbed up

Figure 20 Contrasting textures in adjacent lava clasts within coarse greywackes at Bennane Lea. Plane polarised light, ×20 S 69698, PMS 471

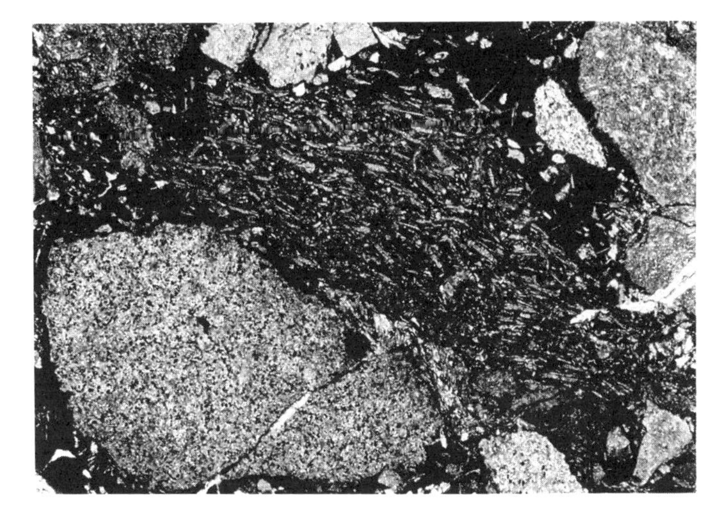

sequence (southward) as a result of both slumping and tectonism, the latter having produced a series of tight minor folds congruous with the proposed larger antiform. The minor folds are well exposed in the cliffs south of Bennane Cave (Figure 16) where hinges plunge at about 50° to the west, a disposition which causes the highest stratigraphical levels seen to be exposed on the foreshore at Bennane Lea. These youngest rocks, contained in the axial zones of the minor synclines, are coarse, probably mass-flow deposits, of greywacke and conglomerate interbedded with slumped cherts. The clasts in the greywackes (Figure 20) and the rounded cobbles in the conglomerate are principally of basalt and chert with some greywacke, massive pyrite and carbonate rock (Bluck, 1978). The carbonate rock is probably a highly altered serpentinite since it contains grains of chrome spinel (Bailey and McCallien, 1957). A very large compound block of brecciated dolerite and clastic tuff marks the southern end of the Bennane Head volcano-sedimentary section. The lower contact of the block with underlying chert appears to be conformable and, since there is no indication of either chilling or contact metamorphism, it may be a large slumped mass. An extreme extention of this possibility was proposed by Dixon (1980) who regarded all of the conglomerate and chert sequence as blocks within a mélange deposit. Despite disagreement over the origin and detailed structure of the sequence there is, however, a consensus that the Bennane Lea chert-conglomerate unit lies at the stratigraphic top of the Bennane Head section.

There is a fair degree of unanimity on the origin of the volcanic portion of the Bennane Head sequence as oceanic island, 'hot spot' basalts. Wilkinson and Cann (1974) on the evidence of Zr-Ti-Y ratios placed the northern (and eastern) margin of the Bennane Head sequence at the major north–south fault to the immediate west of Balcreuchan Port (Figure 16). The lavas and lava breccias of Balcreuchan Port itself they related to the Games Loup, probable island-arc succession. The Bennane Head basalts were at first considered by Lewis (1975) to have an island-arc origin based on major element ratios. However, after further consideration of the rare-earth elements Lewis and Bloxam (1977) were equivocal between island-arc and oceanic island origins, finally favouring the former on the erroneous assumption of an anomalously thick volcanic succession. Jones (1977) confirmed Wilkinson and Cann's Ti-Zr-Y ratio deduction of an oceanic-island origin and more recently Thirlwall and Bluck (1984) described the sequence as consisting of ocean-island tholeiites in terms of both chemistry and Nd-Sr isotopes. An oceanic island or 'hot spot' origin for the Bennane Head volcanics therefore seems well established.

The Balcreuchan Group: southern outcrop

The southernmost outcrop of the Balcreuchan Group intervenes between the southern serpentinite belt and the Stinchar Valley Fault. The outcrop width is substantial in both the north-east and the south-west (Figure 11) but narrows to less than 150 m in its central part south-west of Colmonell. This outcrop pattern is structurally controlled, the volcano-sedimentary sequence being enclosed within an anastomosing fault system trending NE-SW, parallel to the Stinchar Valley Fault. Marked differences in the nature of the sequence in different parts of the outcrop area emphasise the importance of the faulting. Thus, north-east of the narrowest part of the outcrop at Colmonell, the basic lavas of Aldons Hill and the distinctive acidic lithologies of the Craig Hill Breccia Formation stand in marked contrast to the basaltic breccia sequences cropping out south-west of Colmonell in the vicinities of Knockdolian, Sallochan Hill and Corseclays. Farther south-west the unique pyroxene-rich volcaniclastic rocks of the Mains Hill Agglomerate Formation distinguish a third variant.

The Craig Hill to Aldons Hill area: lava sequence

The largest continuous outcrop of lava in the Ballantrae Complex, and yet also the most poorly exposed, lies in an 8 km long belt north-east of Colmonell forming Craig Hill [NX 165 875], Poundland, Glessal and Bargain hills and Aldons Hill [NX 190 900]. Convincing evidence for the attitude of the lavas is sparse; a NE–SW strike with bedding dipping steeply to the south-east is to be seen in the Craig Hill area, whereas the outcrop of a thin feldsparphyric lava on the north side of Fell Hill [NX 1865 9010] indicates a moderate dip to the north-west. A markedly different trend, with a NW–SE strike and a moderate dip to the north-east is locally apparent in the Aldons Hill area, where the lavas appear to be folded into a large open anticline. The lavas are generally crudely pillowed spilitic basalts, predominantly aphyric with only rare plagioclase and augite phenocrysts, and are variably vesicular. Albitised plagioclase, chlorite and augite are the commonest constituent minerals with secondary prehnite, pumpellyite, actinolite and sphene as widespread accessories placing the lava sequence in the prehnite-pumpellyite facies of burial metamorphism (Figure 13; Smellie, 1984a). Extensive silicification has affected the basalt lavas in a broad zone north of Fell Hill, becoming more patchily developed south-east towards Aldons Hill, to produce grey-green flinty rocks.

Chemical evidence for the eruptive setting of the lavas is ambiguous. Lewis (1975) and Lewis and Bloxam (1977) regarded the Craig–Aldons sequence as the oldest part of the Balcreuchan Group and placed it at the base of a thick island-arc low-K tholeiitic lava pile which originally continued upwards into the Knockdaw Hill succession. However, their major-element and rare-earth-element data were equivocal between island-arc and oceanic-ridge environments, and they preferred the island-arc environment on the basis of other features of the sequence, some of which have subsequently been reinterpreted. An alternative ocean-ridge origin for the Craig–Aldons basalts was favoured by Jones (1977) on the basis of Ti-Zr-Y ratios.

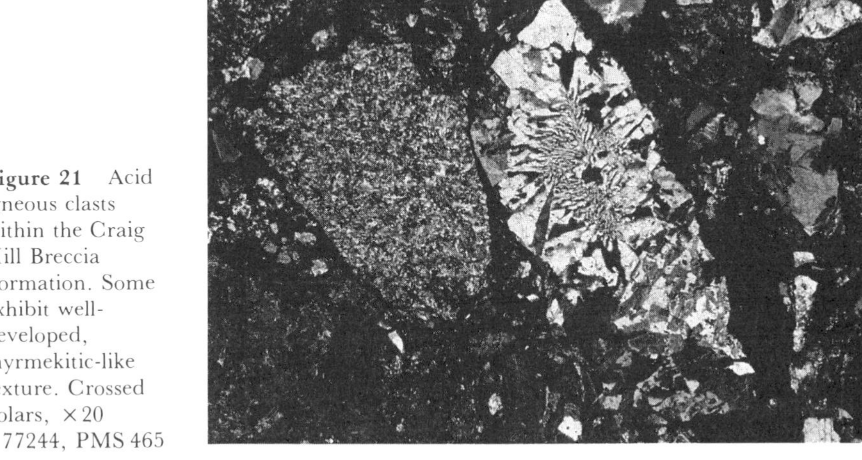

Figure 21 Acid igneous clasts within the Craig Hill Breccia Formation. Some exhibit well-developed, myrmekitic-like texture. Crossed polars, ×20 S 77244, PMS 465

The Craig Hill to Aldons Hill area: Craig Hill Breccia Formation

The proportion of basaltic breccia and interbedded sedimentary rock in the lava sequence is generally low, with the exception of the Craig Hill/Poundland Hill area, where the Craig Hill Breccia Formation and several chert horizons crop out. The breccia is unique in the Balcreuchan Group in containing coarse-grained acid igneous rocks (Figure 21, S 57758, 77244) among the subangular to rounded clasts. Basalt, both aphyric and feldsparphyric, is the commonest clast type, but the full range includes dolerite, siltstone, chert, carbonate, amphibolite, feldspar, chrome spinel, quartz, myrmekitic and graphic granite, hornblende-riebeckite granite, microgranite and quartz diorite. Rarely the clasts, which range up to 1.5 m in diameter, are supported in a mélange-like sheared mudstone, but generally the matrix, usually chloritic but silty and even sandy locally, is sparse. Balsillie (1937) and Jones (1977) considered the breccia to be the result of in-situ brecciation and silicification, whereas Lewis (1975) described it as agglomerate or possibly an intrusion breccia, interlayered with tuff and chert. Stratification is well developed along the southern margin of the outcrop in Craig Wood [NX 165 869] and is approximately parallel both to the bedding attitude of cherts exposed nearby and to the overall trend of layering in the pillow lavas. The only exposed local source for the coarse-grained acid igneous rocks is the Byne Hill trondhjemite body, from which some comparable lithologies have been described (Bloxam, 1968). If the Byne Hill body is accepted as the source then the breccias must be one of the younger units in the Ballantrae Complex, younger than the 483 ± 4 Ma U-Pb date of the trondhjemite (Bluck and others, 1980). The Craig Hill Breccia Formation appears to be conformably interbedded with the basalt lava sequence; despite the low proportion of matrix, the mélange-like nature of a small part of the formation suggests slumping as the most likely mode of origin.

Knockdolian, Sallochan Hill and Corseclays

The coarse basalt breccias well exposed on the slopes and summit of Knockdolian were thought by both Geikie (1897) and Peach and Horne (1899) to be agglomerates deposited in close proximity to a volcanic vent. Peach and Horne commented on the lack of matrix and the angularity of the clasts in the breccia, which they considered overlay aphyric lavas on the north-east side of the hill. The aphyric lava certainly overlies feldsparphyric lavas with interbedded sandy horizons, the whole sequence being folded into a broad, open anticline with axis plunging to the west-south-west. There is a marked similarity between this conformable upward sequence of feldsparphyric lava to aphyric lava to breccia, and the proven succession in the Bennane Head area (Stone and Rushton, 1983).

The Knockdolian breccias are formed entirely of basalt clasts which are usually angular but range to subrounded. The general clast size is less than 10 cm in diameter, but blocks up to 1 m across are included, particularly on the western side of the hill (possibly the stratigraphically highest beds), where the breccias tend to be coarser and include a higher proportion of subrounded fragments. The clasts are entirely basaltic and, with the exception of sparse and small phenocrysts of plagioclase and/or augite, are aphyric consisting mainly of fine-grained intergrowths of plagioclase and augite. Many of the clasts are vesicular. Petrographically all of the clasts are identical to the supposedly underlying aphyric lavas and to the matrix of the feldsparphyric lavas, the phenocrysts of which are albitised plagioclase. The presence of abundant secondary chlorite in the lavas together with accessory epidote, sphene and rare prehnite, coupled with the albitisation of the plagioclase, suggests that this part of the sequence has been affected by prehnite-pumpellyite-grade burial metamorphism (Figure 13). By contrast the apparently overlying breccias contain smectite in excess of chlorite and more calcic, less albitised, plagioclase: an assemblage more indicative of a lower-grade, zeolite-facies metamorphism (Smellie, 1984a). This distribution may support the stratigraphic scheme of Peach and Horne (1899), but elsewhere, to the east of the feldsparphyric lava outcrop, aphyric basalts are exposed with a probable zeolite facies secondary mineral assemblage. Thus the metamorphic relationships are not entirely clear, and the facies distribution may be the result of unrecognised structural imbrication such as is seen at Bennane Head (Smellie, 1984a; Stone and Rushton, 1983; Stone, 1984).

The stratigraphical and structural deductions of Peach and Horne (1899) were broadly supported by Lewis (1975) and by Lewis and Bloxam (1980) in an account of the mechanism of formation of the 'macadam' breccias. By contrast, Jones (1977) dismissed the anticlinal structure and described the whole sequence as dipping generally westward at angles of between 45° and 60°, a structural interpretation difficult to reconcile with the admittedly sparse field evidence. From the Ti-Zr-Y ratios of the lavas Jones deduced an origin for the sequence in an oceanic, Hawaiian-type, island environment. However, Lewis (1975) considered that the major-element ratios of the lavas were compatible with their origin as island-arc quartz tholeiites.

All recent investigations agree that the breccias are an essentially water-laid sequence of coarse clastic rocks exceeding 500 m in thickness and produced by the autoclastic disintegration of aphyric pillow lavas rather than as the explosive agglomerate envisaged by Peach and Horne. The presence of some subrounded clasts argues for an input of debris from a fairly local source subjected to wave action. A submarine origin is also evinced by the rare black shales and cherts interbedded with the lavas.

The Sallochan Hill outcrop is dominated by breccias very similar to those of Knockdolian but of generally smaller clast size, with individual subangular to subrounded fragments usually less than 4 cm in diameter. Some well rounded clasts are also present, and they are far more abundant than in the Knockdolian breccias. There are also thin, interbedded lavas within the Sallochan breccia sequence and rather more chert horizons, both black and red, than crop out on Knockdolian. These interbedded sedimentary horizons define a variable north-eastward strike with an invariably steep dip. The lavas and breccia clasts are all aphyric basalts consisting essentially of plagioclase and augite with accessory and secondary smectite, sphene and chlorite. The plagioclase is partially but by no means completely albitised, which together with the secondary mineral assemblage suggests burial metamorphism in the zeolite facies. The Sallochan Hill breccias are thus very similar to the adjacent breccia sequence on Knockdolian, to which they are probably closely related. The two sequences are now separated by a major fault zone followed by the present valley of the River Stinchar.

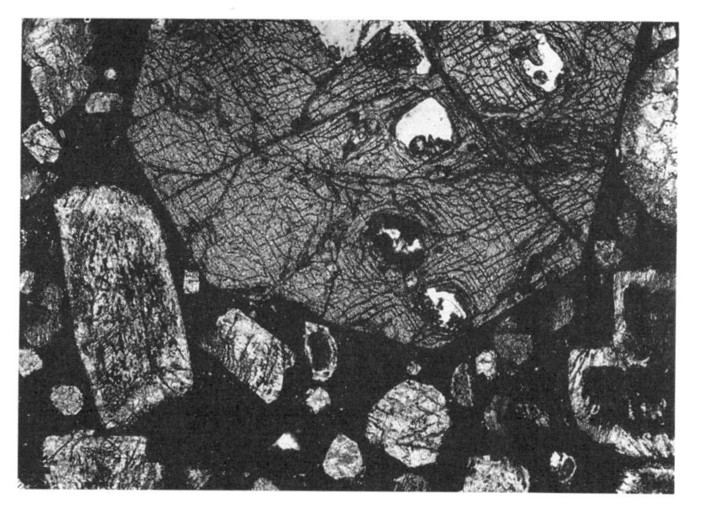

Figure 22 Clinopyroxene and feldspar phenocrysts in basalt of the Mains Hill Agglomerate Formation. Plane polarised light ×20 S 69707, PMS 460

The very poorly exposed and deeply weathered aphyric and feldsparphyric basalts and basalt breccias exposed in the vicinities of Corseclays and Balig farms are possible correlatives of the Knockdolian-Sallochan Hill rocks. The breccias at Balig contain a few clasts with small augite phenocrysts in addition to the more common and widespread aphyric basalt clasts. At no locality could the bedding attitude be confidently assessed.

The Mains Hill Agglomerate Formation

In a zone extending north-east to south-west between Balnowlart Farm [NX 110 836] and Ballantrae village [NX 085 825], separated from the Sallochan Hill breccia by a major fault, the volcanogenic sequence is markedly different from the rest of the Balcreuchan Group. A characteristic of this formation, first noted by Peach and Horne (1899), is the abundant fresh augite that forms phenocrysts both in lava and in lava clasts in agglomerate (Figure 22), and occurs as discrete clasts in the agglomerates and tuffs. The tuffs form the greater part of the succession, with the lavas occurring as isolated and irregular masses up to 100 m broad within them. The size and distribution of the lava bodies suggested to Peach and Horne that they represented large slumped masses, a possibility reinforced by signs of soft sediment deformation in adjacent tuffs. However, for the most part the succession seems to be truly pyroclastic with a chaotic jumble of angular to subrounded clasts of pyroxene and plagioclase-phyric basalts, broken plagioclase and augite crystals all set in a very fine-grained greenish matrix.

The lava clasts, ranging up to about 15 cm in diameter, and the very much larger, presumed foundered, lava masses are all markedly porphyritic with abundant sericitised plagioclase and fresh euhedral augite phenocrysts varying in proportion but jointly making up 30 – 50 per cent of the rock (Figure 22). The plagioclase is patchily albitic, but chemically the basalts do not seem to have been extensively spilitised (Thirlwall and Bluck, 1984). More rarely phenocrysts of olivine are also present completely pseudomorphed by chlorite or by bowlingite and magnetite. Lewis (1975) speculated that augite accumulation was responsible for the unusual petrography. Prehnite, pumpellyite and actinolite occur as secondary accessories in the Mains Hill area at the western end of the outcrop (Smellie, 1984a) showing that part of the sequence to have been subjected to prehnite-pumpellyite facies burial metamorphism. However, at the eastern, Balnowlart, end of the outcrop prehnite and pumpellyite do not occur but instead the basalt clasts contain zeolite, analcime and smectite indicative of a significantly lower burial metamorphic grade (Figure 13).

Bedding indicators within the agglomerate are obscure in the Balnowlart area but along the southern margin of the outcrop adjacent to the River Stinchar, interbedded unfossiliferous fine-grained tuffs and red chert strike NE–SW and are steeply inclined. Grading and cross lamination in tuffaceous siltstone show that the sequence youngs to the north-west. If this trend were continued across the outcrop it would imply that the stratigraphically higher part of the sequence has been subjected to a higher grade of burial metamorphism than has the lower part. Some structural complication of the sequence is therefore likely, and the apparent stratigraphical succession of approximately 500 m of agglomerates and tuffs, the coarser lithologies towards the top of the sequence and the lower tuffs interbedded with tuffaceous siltstone, red mudstone and chert, may be an oversimplification.

The exceptional, porphyritic nature of the basalts in the Mains Hill Agglomerate Formation causes considerable difficulty in any attempt to assess their chemical affinities. Both Lewis (1975) and Jones (1977) were concerned by this problem and did not extrapolate their interpretations of the Balcreuchan Group's origins into the Mains Hill area, although Lewis clearly favoured a comparison with island-arc tholeiites. More recently, Thirlwall and Bluck (1984) have studied the rare-earth-element abundances and trace-element ratios and have compared the basalts to those of modern mature island arcs. Further, they obtained a Sm-Nd age from the augites of 501 ± 12 Ma which, whilst allowing an error overlap with the lower Arenig Balcreuchan Group sequences elsewhere, also raises the possibility that the Mains Hill Agglomerate Formation is the oldest volcanic unit within the Ballantrae Complex.

Late Arenig to early Llanvirn intrusive igneous components of the Ballantrae Complex

4

The Ballantrae Complex contains numerous igneous masses which, whilst intrusive and cross-cutting with respect to other elements of the Complex are themselves unconformably overlain by Llanvirn and younger sedimentary rocks. Intrusion of these gabbro, dolerite and trondhjemite bodies was clearly a late event in the formation of the Complex.

Minor intrusions in the northern and central areas

Dolerite (e.g. S 67251) is the most common intrusive rock within the northern serpentinite outcrop, forming all of the prominent, upstanding masses on the foreshore between Burnfoot [NX 108 883] and Pinbain Bridge [NX 138 915], and most of the rocky knolls within the serpentinite outcrop inland. The dolerites are grey, weathering brown or green, medium-grained rocks, generally containing numerous small (1–3 mm) feldspar phenocrysts. Some intrusions are feldsparphyric, others are completely aphyric and some become gabbroic locally. In addition to the feldspar they contain colourless augite or purple titanaugite, and accessory hornblende, biotite, ilmenite and apatite (Balsillie, 1932). They vary from steeply dipping sheets to podiform or lenticular masses whilst individual intrusions sometimes comprise several pod-like masses arranged *en train*, with smooth, dome-shaped original chilled surfaces projecting above the erosion level of the enclosing serpentinite. Thicknesses vary from about 10 m down to thin sinuous stringers a few centimetres across. Chilled margins are everywhere evident in the form of prominent, flinty and porcellaneous rims, usually 1–3 cm thick, truncated sharply against dark-coloured serpentinite. In places, a thin baked rind of serpentinite adheres to the chilled dolerite, and there are rare serpentinite enclaves. Cross-cutting relationships between intrusions are not uncommon and measured orientations box the compass. However, the dykes generally have a bimodal distribution, trending predominantly to the NE and ENE (Figure 23; cf. Peach and Horne, 1899; Balsillie, 1932).

Although no dolerite intrusions cross-cut the north-western faulted margin of the serpentinite outcrop, several (S 67288, 67586) cross the south-eastern margin into the adjacent lavas and volcaniclastic rocks, clearly truncating the

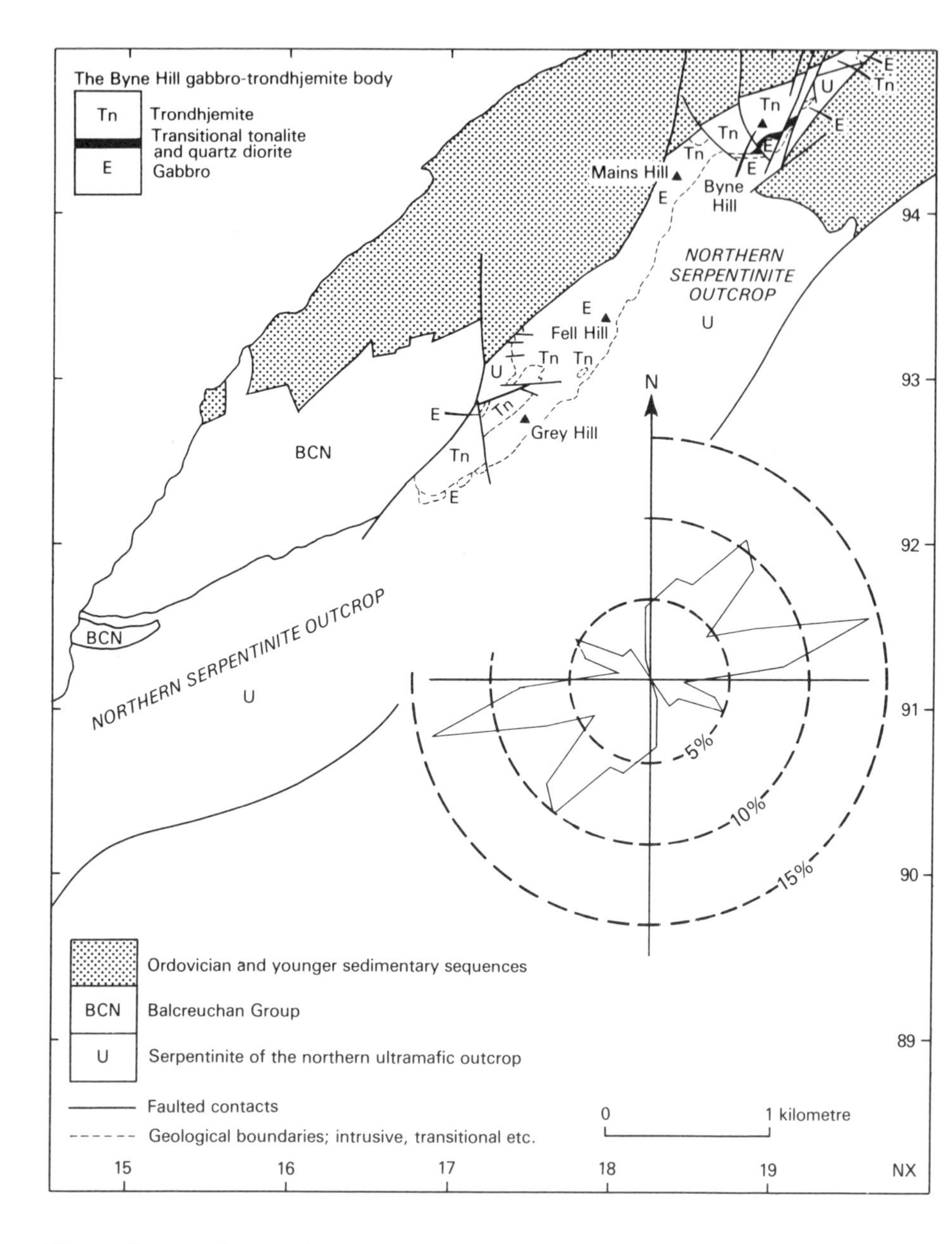

Figure 23 Location and internal composition of the Byne Hill gabbro-trondhjemite body together with a summary diagram of the trends of 94 late Arenig–early Llanvirn dykes from the northern serpentinite outcrop

foliation in the intervening metamorphic sole. This is particularly well seen on the north-western slopes of Balsalloch and Carleton hills and in the nearby Lendal Water at Straid Bridge, and 500 m north-west of Laigh Knocklaugh [NX 1689 9217]. Beyond the south-eastern margin of the northern serpentinite outcrop, broadly similar gabbroic rocks intrude lavas and sedimentary rocks of the Balcreuchan Group between Troax [NX 110 876] and Knockormal [NX 138 887], and the ultramafic outcrop at Balsalloch [NX 116 884]. The gabbros (e.g. S 67268, 67269) generally preserve an ophitic texture between saussuritised plagioclase and diopsidic clinopyroxene which is patchily replaced by amphibole. Serpentine and chlorite pseudomorphs after olivine are common accessories.

Some of the more southerly gabbro bodies, such as that at Knockormal (S 67278, 67598, 67608), are almost entirely fault-bounded but elsewhere the fault zone itself may be cross-cut by gabbro, for example (S 67257) at Burnfoot [NX 108 883]. Since there is no evidence to suggest the fault-bounded gabbro was intruded into an existing fault zone, and given the widespread shearing and brecciation in the Knockormal gabbro, it is likely that there are at least two phases of gabbro intrusion separated by a period of faulting. A two-stage intrusion history has also been deduced from geochemical data for the dolerite dykes cutting serpentinite between Burnfoot and Pinbain Bridge by Holub and others (1984). They divide the dykes into an earlier group with the chemical characteristics of island-arc tholeiites and a later, numerically dominant group of alkali dolerites with the chemical characteristics of within-plate basalts. The earlier, island-arc dolerites have clinopyroxene extensively altered to

Figure 24
Banded gabbro contained as a tectonic inclusion in the northern serpentinite belt at Whilk. D 3338

actinolitic hornblende whereas ophitic textures involving little-altered augite are common in the later alkali dolerites.

Two distinct groups of basic igneous rocks are also seen within the northern serpentinite adjacent to its south-west margin at Whilk [NX 118 890]. The relationships there are confused but it seems likely that at least some of the basic bodies are tectonic inclusions within the serpentinite. These include banded gabbros (Figure 24) which are not seen in situ elsewhere within the Ballantrae Complex but which are a crucial element in ophiolite 'pseudostratigraphy'. Their presence is therefore vital to any interpretation of the Complex as a dismembered ophiolite (Church and Gayer, 1973; Bluck, 1978).

The trondhjemite suite

The largest of the late intrusions forms the north-eastward-trending ridge between Grey Hill [NX 1646 9280] and Byne Hill [NX 1810 9487]. It consists of an elongate dome with steeply dipping sides, cored by trondhjemite which merges, through a narrow transitional zone of tonalite and quartz diorite, into dolerite and gabbro (Bloxam, 1968). Perhaps significantly the elongation parallels the dominant dyke trend (Figure 23). All contacts are gradational except for the outer margin, where gabbro and dolerite are clearly chilled against serpentinite.

Trondhjemite (S 71048–71050, 71052, 71053) forms most of Byne Hill and large parts of Grey Hill. It is a pale pink or orange coloured rock, which weathers white, and is largely formed of albitic plagioclase and quartz. These minerals together comprise about 90–95% of the rock, with accessory dark green hornblende, biotite and zircon. The grain size is generally 1–2 mm, ranging up to 3 mm in places. The marginal basic rocks are mainly dark grey dolerites on Grey Hill but become progressively more gabbroic northward towards Byne Hill. Parts of the gabbro (S 71047, 71051) are pegmatitic, with individual crystals up to 5 cm across. Neither planar nor layered structures are present. Passing in towards the trondhjemite, the lithology changes from olivine gabbro to hornblende gabbro by gradual loss of olivine and progressive replacement of clinopyroxene by brown hornblende. The contact between gabbro (or dolerite) and trondhjemite is not clearly exposed on Grey Hill, but on Byne Hill the two lithologies are separated by a thin zone of pale grey diorite and quartz diorite, which varies in thickness from 1 to 10 m; small patches of this lithology also occur infrequently within the gabbro and trondhjemite. These transitional rocks contain distinctively lath-shaped albitic plagioclase, biotite, brown and green hornblende, and traces of quartz and apatite.

Zircons from the trondhjemite have given a U-Pb date of 483 ± 4 Ma (Bluck and others, 1980) similar to the age (478 ± 8 Ma) of the metamorphic aureole marginal to the northern serpentinite belt, itself cut by other members of the intrusive suite. However the chilling of the basic margin of the trondhjemite against ultramafic rock suggests that the ultramafic rock was cool at the time of intrusion, thus, intrusion is likely to be a subsequent event to the formation of the metamorphic aureole by the same ultramafic mass when hot.

Gabbro intrusions in the southern area

Within the southern serpentinite outcrop a small number of in-situ intrusions occur, mainly in the south-west of the outcrop north of Knockdolian at Duniewick [NX 116 852] and south-east of Sallochan Hill close to Knockdhu Bridge [NX 132 845]. In both places ophitic gabbro (S 68442) is chilled against the serpentinite and has not been subjected to the same degree of low-grade metamorphism as have the hornblende hornfels xenoliths in the same area. Sericitised and saussuritised plagioclase and a colourless clinopyroxene, probably augite, are the dominant components with secondary chlorite a ubiquitous accessory. Some limited alteration of the augite to secondary hornblende has also occurred and is best developed in the Knockdhu Bridge gabbro. A similar lithology, but with the addition of small serpentine and chlorite pseudomorphs after olivine and very little secondary amphibole, crops out as a fault-bounded strip between Balnowlart [NX 112 840] and Mill Hill near Ballantrae [NX 091 830]. No chilling relationships are preserved and this gabbro (S 68440) is tectonically isolated from the adjacent serpentinite to the north and the Mains Hill Agglomerate Formation to the south. However, it is petrographically identical to, and considered coeval with, gabbro (S 69720–69723) which intrudes and is chilled against the Mains Hill agglomerates at Ardstinchar Castle near Ballantrae [NX 087 824]. In the Ardstinchar Castle gabbro (S 68480, 69711) saussuritised and sericitised plagioclase has a well developed ophitic relationship with augite. The augite is in places rimmed by chlorite which, together with serpentine, also forms pseudomorphs after accessory olivine. Skeletal ilmenite and apatite are also present as common accessory minerals.

The most westerly gabbro outcrop in the Ballantrae area forms the foundation of Ballantrae harbour wall [NX 080 830]. The lithology is an ophitic gabbro (S 68446–68448) with altered plagioclase and augite accompanied by accessory ilmenite, apatite and secondary chlorite. No chilling is evident since all of the exposed margins of the gabbro are faulted against Permian breccias. This faulting is also reflected in cataclastic zones up to 5 cm broad which anastomose through the gabbro. Because of its present structural situation the original relationships of this gabbro are dubious, but it is tentatively associated with the late Arenig intrusive suite.

Rodingitisation

Most of the lower Ordovician minor intrusions which are enclosed by ultramafic rock are conspicuously veined by white, fibrous pectolite and/or prehnite. The veining is genetically related to an alteration process which has particularly affected the chilled, marginal parts of the intrusions resulting in a pale grey or white coloured, very tough rock with a sharp flinty fracture. The best development of the veining and alteration occurs in the doleritic minor intrusions within the northern ultramafic outcrop (e.g. S 67252, 67267, 77242). To a lesser extent similar effects occur within many of the pyroxenites in the ultramafic sequence and the beerbachite and hornblende hornfels xenoliths in the southern serpentinite outcrop (e.g. S 69687).

The alteration, found exclusively in rocks within, or adjacent to, serpentinite is a form of contact reaction (Coleman, 1967) in which the primary igneous minerals are partially or completely transformed into a secondary, Ca-rich mineral assemblage. The resulting calc-silicates or Ca-metasomatites were termed rodingites by Bloxam (1954, 1964). The essential alteration minerals are brown isotropic hydrogarnet (rich in grossular and andradite), prehnite, pectolite, chlorite (mainly penninite), idocrase, diopside, tremolite/actinolite and epidote (pistacite); brown microgranular pumpellyite is a rare accessory and there is localised veining by fibrous serpentine. On a microscopic scale, the secondary minerals largely form pseudomorphs after the primary phases, but some element mobility is evident in the form of secondary veinlets cross-cutting the original igneous grain boundaries. Many of these veinlets follow faulted offset boundaries of the dolerite, and their participation in marginal cataclased contact zones attests to some metasomatism accompanying localised deformation. The associated serpentinites exhibit few obvious textural or mineralogical changes, although Jones (1977) noted a thin chloritic 'blackwall' alteration along the contacts and Bloxam (1954) recorded veins of diopside-zoisite in serpentinite.

The metasomatic process involved enrichment in calcium, and depletion in alkalis and silica; some other major oxides (e.g. MgO, Al_2O_3) show inconsistent trends (Jones, 1977). The source of the calcium is uncertain. During serpentinisation, Ca in orthopyroxene and clinopyroxene may be released and become concentrated in the residual fluid.

Although amounts of Ca released by this process are small it has been proposed (Coleman and Keith, 1971) that serpentinisation of harzburgite and dunite can produce an excess of Ca sufficient for rodingitisation. Inflowing meteoric or connate water released during dewatering or burial metamorphism of underlying lavas or volcaniclastic sediments was possibly an additional source of Ca (Malpas, 1979; Searle and Malpas, 1980) to that produced by the serpentinisation of the pyroxene.

The desilication seen in the rodingites can also be linked with serpentinisation. When serpentine forms from olivine, if it is assumed that no brucite ($Mg(OH)_2$) is developed, there is a shortfall of silica required to balance the reaction. The serpentinisation of orthopyroxene produces excess silica, but it would require 40 – 50 wt per cent orthopyroxene in harzburgite to create serpentinite without an external silica source (Coleman and Keith, 1971; Coleman, 1977). The Ballantrae serpentinites originally contained, on average, only about 18 wt per cent orthopyroxene (Jones, 1977) and so the desilication in the rodingites suggests that they provided such an external source and indicates a genetic link between serpentinisation and metasomatism. If serpentinisation and rodingitisation are coeval and complementary processes, they must have occurred under essentially the same physical and chemical conditions. The bulk of the Ballantrae Complex serpentinite is formed of lizardite and/or chrysotile, which can form in the range 350°C down to ambient temperatures (Barnes and O'Neil, 1969) and, indeed, most serpentinites appear to have formed in the temperature range 100 – 300°C (Wenner and Taylor, 1971). This is consistent with the mineralogy of the rodingites, which is dominated by minerals characteristic of low-grade metamorphism. The lack of pressure-dependent index minerals makes any useful estimate of the pressure impossible (Jones, 1977).

The Downan Point Lava Formation 5

On the south side of the River Stinchar an outcrop of dominantly aphyric pillow lavas extends from the Dalreoch area [NX 160 857] to the coast broadening south-westward. It is bounded to the north by a branch of the Stinchar Valley Fault and to the south by the Dove Cove Fault (Figure 11) of Leggett and others (1982). Exposure is poor in the area of the map but the coastal exposures to the south-west in the vicinity of Downan Point [NX 070 806] are justifiably renowned for their extensive spread of classical pillow-structured basalts (Figure 25). This volcanogenic sequence was originally regarded as an integral part of the Ballantrae Complex (e.g. by Peach and Horne, 1899; Bailey and McCallien, 1952; Bloxam, 1960) but Walton (1961) drew attention to the interbedding of pillow lavas with the Llanvirn–Caradoc Tappins Group, thus introducing the possibility that the geographically adjacent Downan Point Lava Formation was also significantly younger than the Arenig Balcreuchan Group and not a part of it. The field evidence for this interbedding (exposed beyond the margins of the map in the vicinities of Barr [NX 275 940], Currarie [NX 060 780] and Portandea [NX 047 755]) is equivocal and no definitive age data for the Formation has been forthcoming, either biostratigraphic or radiometric. However, there is at present a growing acceptance that the Stinchar Valley Fault is the true western extension of the Southern Upland Fault (Leggett and others, 1979) and therefore the logical southern margin of the Ballantrae Complex *sensu stricto*. It is possible that the base of the Downan Point Lava Formation may extend into the Arenig to correlate with the oldest known spilitic basalts in the Southern Uplands, which are exposed at Raven Gill near Abington (Lamont and Lindström, 1957) 95 km to the north-east, approximately along strike. Certainly if an imbricate thrust model for the Southern Uplands (Leggett and others, 1979; Stone and others, in press), is extended to include the Downan Point Lava Formation then it would be expected that the lavas are contained in the oldest 'slice' of the thrust stack (Barrett and others, 1981 and 1982).

Petrographically the Downan Point Lava Formation consists of tholeiitic pillow basalts which are predominantly fine grained, aphyric and vesicular. Plagioclase, clinopyroxene

Figure 25 Well developed basalt pillows within the Downan Point Lava Formation near Sgavoch Rock. The pillows, which show marked vesicle zonation, range from about 20 cm to 1.5 m in length. D 1572

and chlorite form the matrix, which contains rare, but very locally abundant, small plagioclase phenocrysts (S 69692, 69693). The vesicles, which occur in marked zones parallel to the pillow margins, may be filled by chlorite or calcite. Locally the basalts are interbedded with breccias (S 69694), finely banded tuff and thin horizons of chert. These, together with the pillow attitudes, define a steeply dipping or vertical succession striking NE–SW with a consistent north-westerly sense of younging. Pillow shape is exceptionally well developed (Figure 25; cf. Peach and Horne, 1899; Bloxam, 1960) and ranges from slightly elliptical with long axis of approximately 20–50 cm through to larger, elongate 'bolster' shapes possibly formed in lava tunnels. These in turn merge with interlayered and unpillowed sheet flows. The interstices between the individual pillows are locally filled either by chert or carbonate, some of which is spectacularly crystalline.

From a consideration of aspects of the geochemistry of basalts from the Downan Point area several authors have described an oceanic-island origin for the Downan Point Lava Formation basalts. Wilkinson and Cann (1974) studied the Ti-Zr-Y-Nb ratios and compared the sequence to 'tholeiitic hot spot basalts', whilst Lewis and Bloxam (1977) considered the rare-earth element abundances to be similar to 'Hawaiian tholeiites'. Thirlwall and Bluck (1984) produced analytical data similar to that of Lewis and Bloxam and agreed on an ocean-island environment of eruption.

Within the area covered by the 1:25 000 sheet exposure is limited and the Downan Point outcrop is extrapolated north-eastwards on sparse data. The best available sections in the inland part of the outcrop, within the map area, are in the banks of the River Stinchar at Heronsford [NX 121 835], in the Craigneil Burn [NX 149 853] and in Pyet Glen [NX 152 853].

Sedimentary sequences of middle to late Ordovician age 6

The Llanvirn to Ashgill sedimentary sequences cropping out in the map area are divided between four groups (Figure 26). Unconformably overlying the Ballantrae Complex is the Barr Group, composed dominantly of conglomerates and limestone of Llanvirn and Llandeilo age, itself then overlain, with local unconformity, by the lithologically diverse Ardmillan Group of Caradoc and Ashgill age. The outcrop of

Figure 26 Outcrop distribution of the Ordovician and younger sedimentary sequences adjacent to the Ballantrae Complex

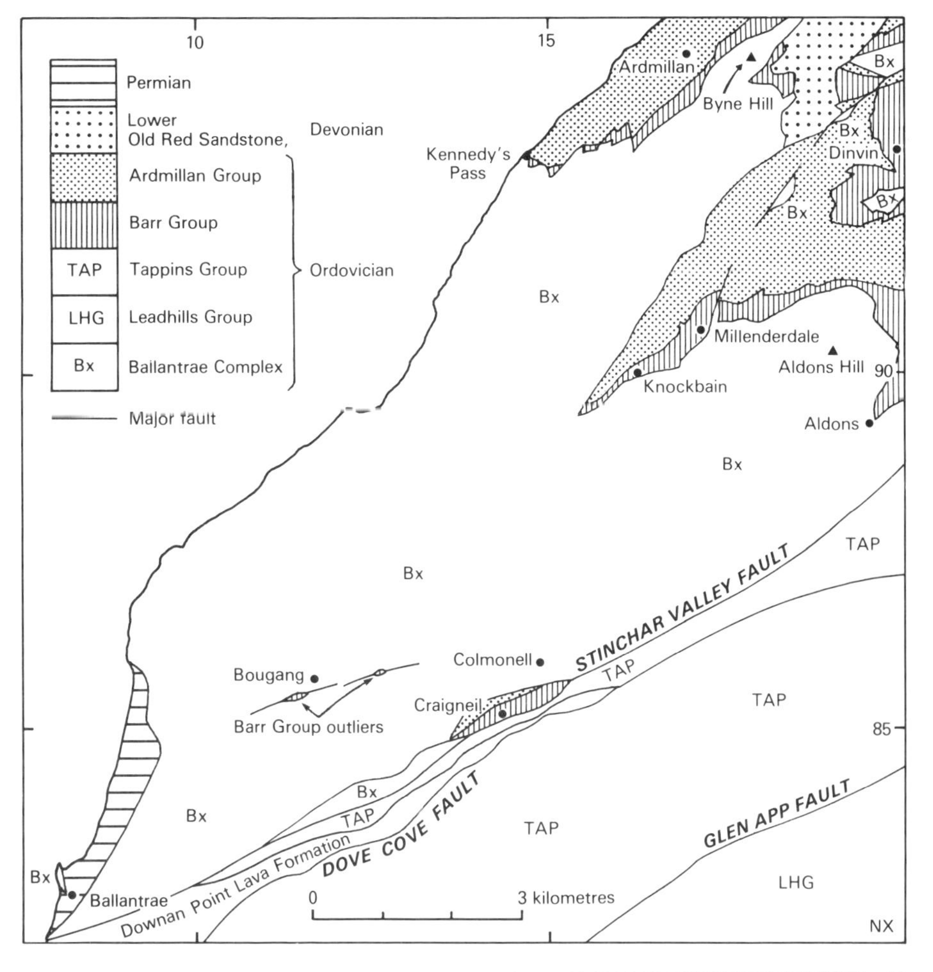

both of these groups is restricted to that part of the map area north of the Stinchar Valley Fault. South of the fault, Tappins Group greywacke-turbidite sequences of possible Llanvirn to Caradoc age are separated by the Glen App Fault from a lithologically similar succession of Llandeilo to Ashgill age. This is a portion of the Leadhills Group; a part of the traditional northern belt of the Southern Uplands. There is therefore a marked lithological contrast across the Stinchar Valley Fault between sedimentary sequences of broadly the same age. With the Barr and Ardmillan groups being deposited broadly proximal to the continental margin, and the Tappins and Leadhills groups turbidites being relatively distal, considerable tectonic shortening is implied in the vicinity of the Stinchar Valley Fault. The present spatial relationship of the groups and their contained formations is summarised in Figure 27.

The Barr Group

Unconformably overlying the Ballantrae Complex and transgressive north-westwards across it (Williams, 1962; Ingham, 1978) is a sedimentary succession dominated by conglomerate with an interbedded, mainly algal limestone horizon. Both Lapworth (1882) and Peach and Horne (1899) recognised and described this tripartite succession as comprising the Kirkland Conglomerate at the base, overlain by the Stinchar Limestone, this in turn overlain by the Benan Conglomerate, the three together forming the Barr Series (Figure 28). Most of the outcrop of this sedimentary assemblage lies in the Barr area several kilometres beyond the eastern margin of the map area, where Williams (1962) refined the earlier stratigraphic scheme and formalised the version shown in Figure 28. Further revision by Ince (1983 and 1984) introduced modern usage of terminology, and his scheme, also summarised in Figure 28, is utilised in this account of the formally renamed Barr Group.

Lapworth (1882) and Peach and Horne (1899) considered the sequence to be Llandeilo in age. Williams (1962) revised this to Caradoc on the basis of graptolites found in his *Superstes* Mudstone, but subsequent revision of Ordovician graptolite biostratigraphy generally has invalidated the change (evidence summarised in Ingham, 1978). The most reliable age indicator at present derives from correlation with the Scandinavian–Appalachian conodont zonation which shows the Llanvirn–Llandeilo boundary to lie approximately in the middle of the Stinchar Limestone Formation (Bergström, 1971). The marked similarity of the brachiopod faunas with Appalachian examples (Williams, 1962) supports the Llanvirn–Llandeilo age and confirms the sequence as part of the Pacific faunal province.

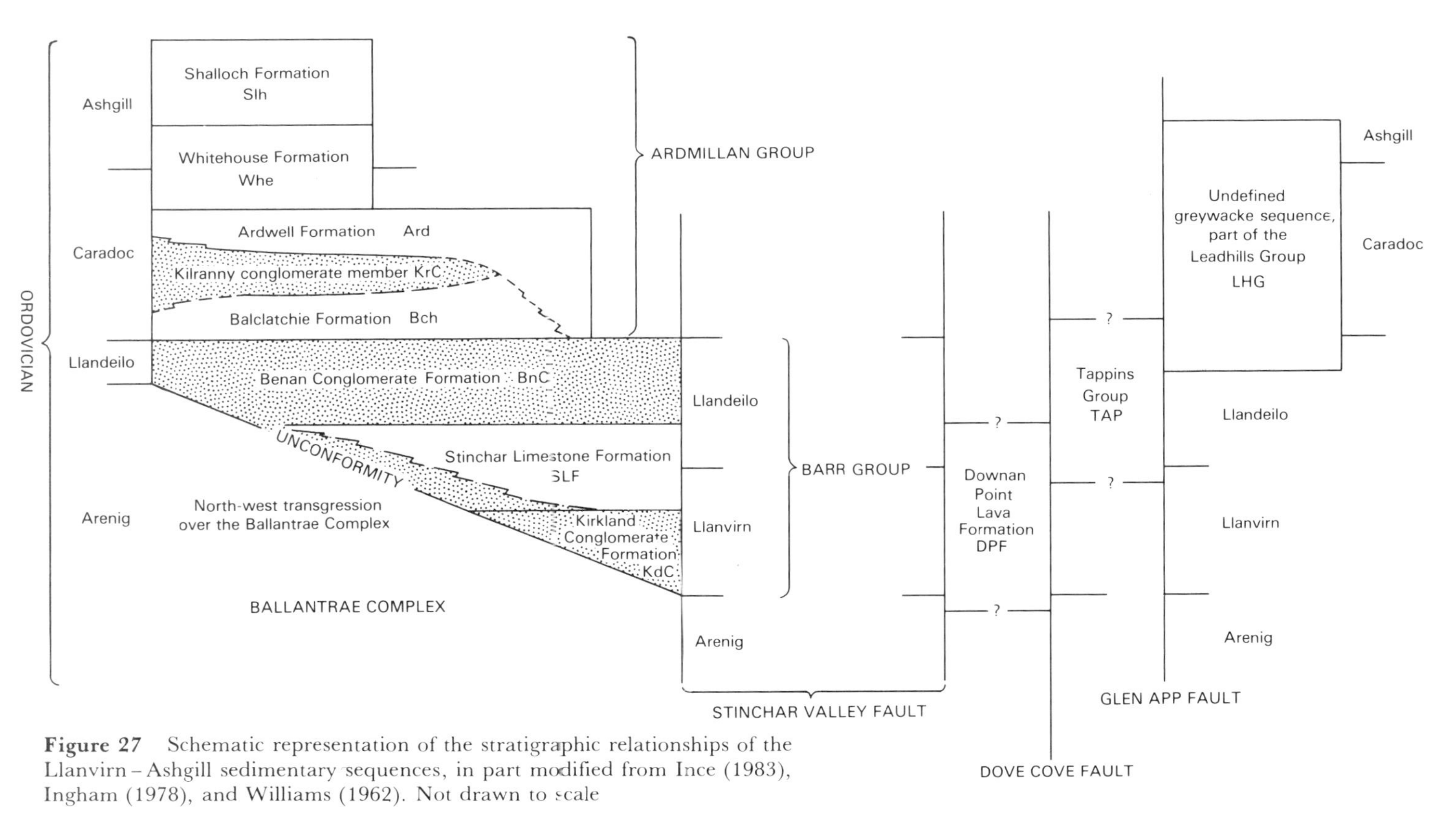

Figure 27 Schematic representation of the stratigraphic relationships of the Llanvirn – Ashgill sedimentary sequences, in part modified from Ince (1983), Ingham (1978), and Williams (1962). Not drawn to scale

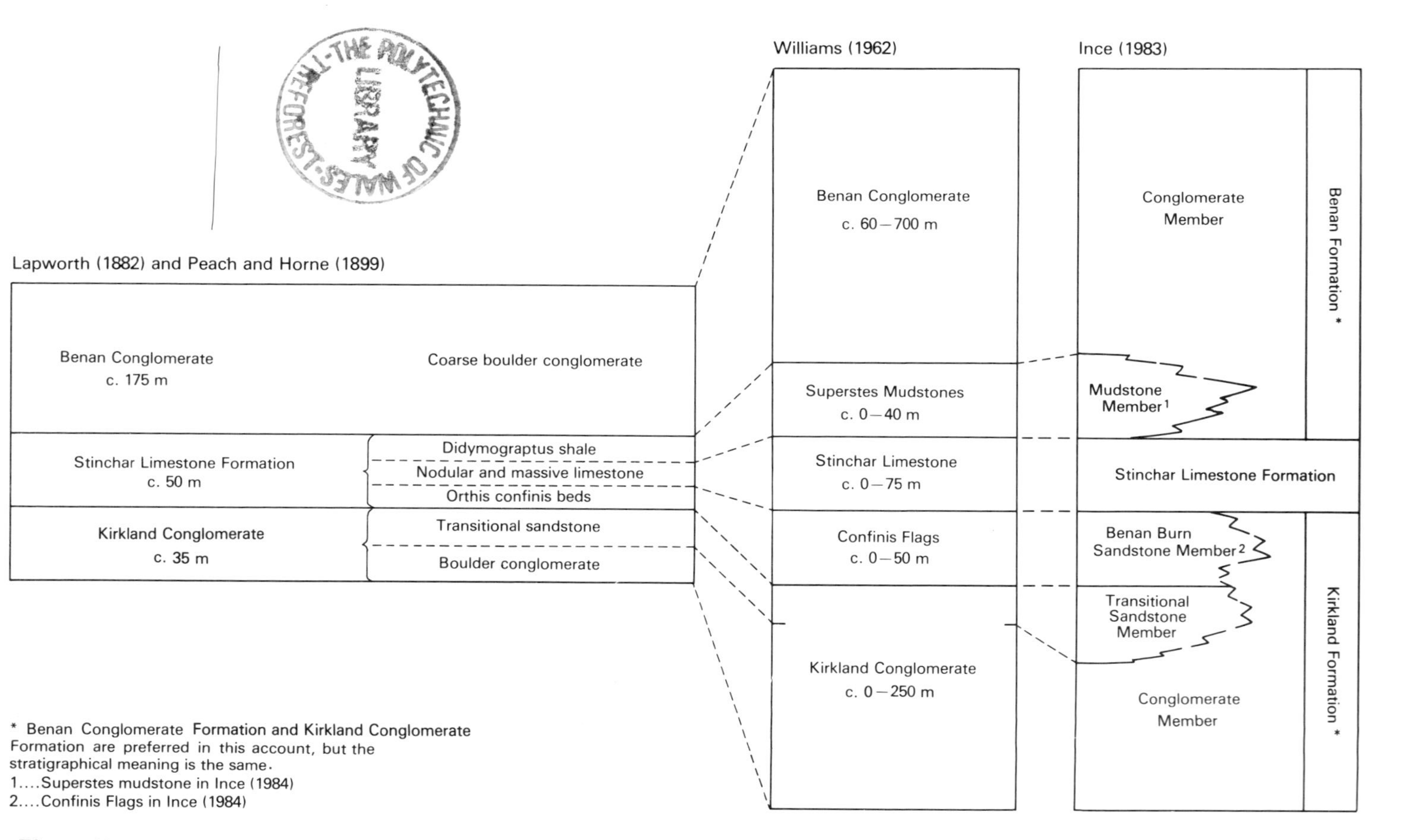

Figure 28 A summary and comparison of the various stratigraphical schemes proposed for the Barr Group

The environment of deposition for the Barr Group is generally accepted (e.g. Williams, 1962; Ingham, 1978; Ince, 1983) to have been a submarine delta or coastal alluvial fan with some of the conglomerates deposited by mass flow. Synsedimentary faulting with downthrow to the south-east probably exercised control on depositional location and thickness. The conglomerates are particularly variable in thickness, but throughout its succession the Barr Group tends to be lithologically inconsistent laterally, which complicates lithostratigraphical correlation.

The basal Kirkland Conglomerate is a clast-supported, pebble conglomerate containing fragments dominantly of basalt, both aphyric and feldsparphyric. Rarer clasts of gabbro, serpentinite and chert are also present together with some pebbles of foliated granite. With the exception of the granite all of the clasts could have been derived from the Ballantrae Complex. An increasing proportion of sandstone is interbedded with the conglomerate in the upper part of the formation and the topmost beds are usually fine-grained sandstone. The overlying Stinchar Limestone Formation is variable laterally, but the commonest lithology is an algal-cemented, rather rubbly limestone with abundant algal debris. Calcareous sandstones are interbedded with the limestone locally.

Above the limestone, a grey-green, silty mudstone is present in some places at the base of the Benan Conglomerate Formation. Elsewhere, boulder conglomerate containing matrix-supported clasts up to a metre across, is erosive into the Stinchar Limestone Formation. The clasts, often well rounded but usually subrounded, are mainly of basalt, dolerite and gabbro but also include vein quartz, chert, greywacke, a very distinctive pink granite and, locally, Stinchar Limestone. Most of the igneous clasts, with the notable exception of the pink granite, could have been derived from the Ballantrae Complex. Exotic clasts of particular importance in terms of palaeogeography are a limestone boulder containing fossils of early Canadian (Tremadoc) age (Rushton and Tripp, 1979) and boulders of unfoliated granite giving Rb-Sr ages of up to approximately 560 Ma (Longman and others, 1979). There is abundant evidence for derivation of the conglomerates from the north-west with the provenance thus including a 'basement' terrain unaffected by the Grampian Orogeny (late Cambrian to early Ordovician), and fringed by a Cambro-Ordovician carbonate platform. Such an assemblage is nowhere exposed and its absence has provoked speculation of major north-westward-directed thrusting or large-scale NE–SW strike-slip movement in the vicinity of the Midland Valley of Scotland during the Ordovician and early Silurian periods (Bluck, 1983; Leggett and others, 1983).

Within the map area the Barr Group sedimentary rocks crop out in three zones (Figure 26): as fault-bounded slices in the Stinchar Valley south and west of Colmonell, unconformably overlying the Ballantrae Complex rocks in a fold system exposed intermittently between Aldons and Knockbain, and in a more extensive outcrop in the north of the map area between Dinvin, Byne Hill and Kennedy's Pass.

The Stinchar Valley outcrops

South of Colmonell the Barr Group crops out over about half a square kilometre in the vicinity of Craigneil Farm [NX 144 853]. The Stinchar Limestone is well exposed in the Craigneil Burn, the adjacent Pyet Glen and in overgrown and partially filled quarries close to Craigneil Farm. Williams (1962) deduced a complex structure for the area involving overfolding and north-westerly-directed thrusting of the Barr Group. The overfolds, with axes trending NE–SW and very steeply inclined axial planes, are thrust over the Caradoc Balclatchie Formation (described below in the section on the Ardmillan Group) and the whole assemblage is contained within anastomosing branches of the Stinchar Valley Fault.

The lowest part of the succession here exposed was recorded by Peach and Horne (1899) and Williams (1962) from the now obscured quarries. There calcareous and variably conglomeratic sandstone containing *Valcourea confinis* was related to the 'Confinis Flags', and may be correlated with the Benan Burn sandstone member (Ince, 1983) of the Kirkland Conglomerate Formation (Figure 28). The sandstones are succeeded by at least 40 m of the Stinchar Limestone Formation, here a nodular, rubbly and sandy limestone containing abundant fragments and in-situ algal growths of *Girvanella*, macrofossil fragments and oolites. Interbedded calcareous sandstone is more abundant here than in the type sections of the Formation near Barr. The top of the limestone is not exposed but the overlying Benan Conglomerate Formation is represented in the Craigneil area by dark mudstone containing limestone nodules and a fauna compared by Tripp and others (1981) to the basal *Superstes* Mudstone of Williams (1962). This is correlated with the basal mudstone member (Ince, 1983) of the Benan Conglomerate Formation (Figure 28).

There are two other outcrops of the Barr Group in the Stinchar Valley area, both very much smaller, and both fault-bounded outliers occurring within the serpentinite outcrop. A small exposure of nodular, rather rubbly limestone interbedded with thin shaley mudstone horizons occurs [NX 126 585] close to Knockdolian Barns, whilst close to Bougang Farm, an outcrop [NX 114 855] of the Stinchar Limestone Formation has been quarried to an extent justifying the construction nearby of now-disused lime kilns. The limestone exposed in the flooded quarry is of the rubbly, nodular variety with mudstone partings and interbedded calcareous sandstones. Bedding is steeply inclined, striking approximately 60° and deduced to young towards the north-west from the evidence of Williams' (1962) description of *confinis* Flags exposed in the south-east corner of the quarry in association with a serpentinite conglomerate. The description suggests that the Bougang outcrop correlates with the top of the Benan Burn sandstone member of the Kirkland Conglomerate Formation (Ince, 1983) and the base of the overlying Stinchar Limestone Formation.

Aldons to Knockbain

Within the map area the best exposures of the Stinchar Limestone Formation are those in two old quarries beside the disused Aldons limeworks [NX 197 896]. The larger (western) quarry was worked in the western limb of a major syncline plunging to the south-west. The thickness of the southern limb of the syncline is reduced by faulting but there too the limestone has been worked in a smaller quarry. At least 18 m of limestone are exposed, most of which is a rubbly *Girvanella*-rich lithology with a variable sand content. A nodular appearance is common and the rock in places consists of subrounded to subangular cobbles of limestone recemented by *Girvanella* algal growths. Some of the sandy horizons contain larger grains and rare granules of basalt and serpentinite which become more abundant down sequence, so that at its base the limestone merges with calcareous pebbly sandstone transitional to a conglomerate of subangular pebbles of basalt in a highly serpentinous matrix. This conglomerate rests unconformably on basalts of the Ballantrae Complex. Above the limestone the basal mudstone member of the Benan Conglomerate Formation contains abundant small carbonate concretions, and passes abruptly up into typically massive, bouldery Benan Conglomerate. To the north of the larger quarry the Benan Conglomerate is erosive into the underlying mudstone and limestone, and around the anticlinal hinge exposed on the slopes of Aldons Hill [NX 195 905] the Benan conglomerate rests unconformably on the Ballantrae Complex. Along the northern limb of the fold some limestone and up to 10 m of dark grey-green mudstone are interbedded with conglomerate between 10 and 15 m above the unconformity over Ballantrae Complex basalts. The mudstone has been correlated with the *Superstes* Mudstone Formation (Ince, 1983), suggesting that the conglomerates underlying the mudstone are the lateral equivalent of the Stinchar Limestone Formation and form a conglomerate facies therein. However, in the Aldons Hill to Knockbain section all of the conglomerates contain the pink granite clasts typical of the Benan Conglomerate (*s.s.*) and the whole unit could perhaps best be regarded as a diachronous transition zone between the Stinchar Limestone and the Benan Formations. The variable development of the succession has been documented in some detail by Williams (1962, pp.18–21) and by Jones (1977, pp.321–331).

Outcrops between Dinvin, Byne Hill and Kennedy's Pass

In the north of the map area the Barr Group has a large arcuate outcrop, cut by numerous faults, between Dinvin [NX 2002 9317] and Kennedy's Pass [NX 1450 9272]. Exposure is generally very poor but the sequence appears to have a faulted or thrust base except around Drumfairn [NX 1807 9403] where 'rounded boulders and cobbles of serpentinite in a sandy matrix (Benan Conglomerate Formation) . . . are found resting unconformably on an irregular serpentinite floor' (Williams, 1962, p.25). Rocks of the Stinchar Limestone Formation are found only at Pinmacher Farm [NX 1956 9361] where conglomerate and limestone form the local base to a major thrust mass (Tormitchel thrust sheet: Williams, 1959). There, Williams (1962, p.18) recorded a sequence of 36.6 m of conglomerate, 2.5 m of dark green sheared mudstone, and 4.6 m of nodular and platy oolitic limestone with an undiagnostic fossil fauna. These exposures are now overgrown except for a small knoll of conglomerate. A 1.5 m section of cream-coloured fractured limestone with numerous thin anastomosing films of dark grey shale was uncovered temporarily in 1983 during drainage operations [NX 1966 9368].

On Dinvin Hill, just east of the sheet boundary, at least 518 m of Benan Conglomerate succeed the *Superstes* mudstones. Thickness decreases to about 183 m on Byne Hill, further reducing to only 18 m on the foreshore south of Kennedy's Pass by a combination of lateral thinning and faulting. The best exposures are at the last two localities where the formation is formed of very crudely stratified, matrix-supported, pebble, cobble and boulder conglomerates with rare, impersistent, thin lenses of sandstone and clast-supported, framework conglomerate.

The Ardmillan Group

The Ardmillan Group, with an age ranging from Caradoc to Ashgill (Ingham, 1978), is dominated by turbidite sandstone, siltstone and mudstone, with conglomerate prominent only in the basal parts of the succession. The formations defined here broadly correspond to stratigraphical units recognised by Lapworth (1882) and Peach and Horne (1899), but incorporate modifications introduced by

Williams (1962) and Ingham (1978). Although the original formation definitions utilised some lithological criteria (especially Lapworth, 1882), the often laterally impersistent nature of the constituent lithologies has resulted in a subsequent reliance on palaeontological criteria: unfortunately lithological and palaeontological boundaries do not necessarily coincide over the entire area. Thus, where palaeontological information is absent, the boundaries between the Balclatchie and Ardwell formations (Williams, 1962), and Ardwell and Whitehouse formations (Ingham, 1978) are difficult to place. For the latter the proposal of Ingham (1978) is here adopted whereby the boundary is placed at the base of the first thick gravelly limestone turbidite of the Whitehouse Formation exposed on the north side of Ardwell Bay. The underlying lithologically contrasting sequence of striped mudstone and thin-bedded fine sandstone is assigned to the Ardwell Formation. This procedure unifies both the lithological and palaeontogical information available. The boundary between the Balclatchie Formation and the Ardwell Formation is a less tractable problem, however, and a Balclatchie-type fossil fauna is known to occur 40 m above the base of the Ardwell sequence, as originally defined (Lapworth, 1882; Williams, 1962). According to the original definition of Lapworth (1882), the Balclatchie Formation corresponds to the sequence of mudstone, siltstone, fine sandstone and conglomerate that is younger than the Benan Conglomerate and older than the basal beds of the Ardwell Flags: the base of the formation is clearly intended as the top of the Benan Conglomerate. However, along the coast north from Kennedy's Pass and inland at least as far as Laggan Hill [NX 2015 9480] the local top of the formation can conveniently be taken as the upper surface of the Kilranny Conglomerate, itself defined as the Kilranny Conglomerate Member of the Balclatchie Formation (Figure 27). This division, based solely on lithology, is close to that originally defined by Lapworth (1882), despite his misidentification of the Kilranny Conglomerate as Benan Conglomerate. In the large area of extremely poor exposure between Knockbain [NX 1626 9002], Pinmacher and High Letterpin [NX 1990 9155], the Kilranny Conglomerate Member is absent, and so, with no basal marker horizon, the Ardwell Formation is continued down to overlie the Barr Group (Figure 27). However the situation is confused by the occurrence of a Balclatchie-type fossil fauna near Knockbain (Williams, 1962) and stratigraphical definition in this area remains inadequate.

There is no evidence from the Ardmillan Group for the very shallow water or possibly even subaerial conditions that recurred during deposition of the underlying Barr Group (Ince 1983, 1984), apart from stromatolitic algal mats reported from the upper parts of the Ardwell Formation (Bluck, *in* Hubert, 1966, 1969, and *in* Ingham, 1978). Their occurrence was not confirmed during the recent resurvey and the entire succession appears to record a period of continuous submarine deposition.

The predominantly thin-bedded, fine sandy and silty sedimentary rocks were probably deposited mainly as low-density turbidites, whereas the interbedded coarse gravelly or conglomeratic beds are probably residual deposits of high-density turbidity currents (cf. Kuenen, 1953; Lowe, 1982). The Kilranny Conglomerate is locally channelled deeply into the underlying Balclatchie mudstones and probably accumulated sequentially by a variety of processes, partly tractional but predominantly mass-flow. Its lithological characteristics are similar to those of the lower parts of the Benan Conglomerate Formation, interpreted by Ince (1983) as resedimented conglomerates accumulated in the distal parts of a fan-delta at outer shelf depths. Such an interpretion of the Kilranny Conglomerate would be consistent with its intercalation between the relatively distal, deep-water turbiditic sequences of the Ardwell Formation and Balclatchie Formation mudstones. However, the depositional environment of the Ardmillan Group as a whole, whether wholly within a neritic, pro-deltaic setting (Hubert, 1966, 1969) or sub-

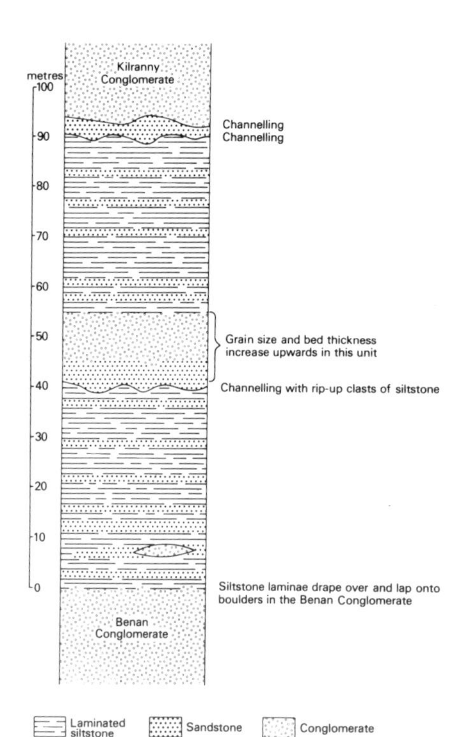

Figure 29 Diagrammatic summary of the lithostratigraphy of the lower Balclatchie Formation as exposed in coastal sections south of Kennedy's Pass

marine fan at outer shelf or slope depths (Ingham, 1978; Anderton and others, 1979) or some combination of these (Walton, 1983), has yet to be resolved.

Balclatchie Formation

The lowermost beds of the Balclatchie Formation are exposed on the foreshore south of Kennedy's Pass and in the Knockbain Burn [NX 164 900]. In Knockbain Burn, Williams (1962, p.25) described approximately 15 m of conglomerate, sandstone and blue-grey silty mudstone yielding trilobites, ostracods and crinoid stems in addition to a diagnostic post-Benan brachiopod fauna (Williams, 1962; Tripp, 1980) overlying the Benan Conglomerate. The complete sequence separating the Benan and Kilranny conglomerates (the infra-Kilranny Greywackes of Williams, 1962) is well exposed on the foreshore south of Kennedy's Pass and is summarised in Figure 29. The basal beds are draped over and locally lap on to boulders and cobbles of the underlying Benan Conglomerate. The sequence then passes up through siltstone, sandstone and subordinate conglomerate until, over a small interval towards its top, this dominantly fine-grained sequence is interbedded with sandy, pebble conglomerates with channelled, erosive bases. These thicken and merge upwards into 120 m of channellised pebbly coarse sandstone containing dispersed cobbles and boulders. The unit becomes wholly conglomeratic locally and develops a weak stratification in its upper part. Together the pebbly coarse sandstone and conglomerate form the Kilranny Conglomerate member.

The base of the Kilranny Conglomerate is best seen in a small quarry on the east side of the coastal road at Kennedy's Pass, where it is erosive into the underlying laminated Balclatchie Formation siltstones to a depth of about 4 m. The conglomerate is well seen in roadside exposures there and on the foreshore, and in the adjacent Bellmoor and Shiel hills, where it attains a general thickness of about 150 m. It consists of grey-green and blue-grey pebbly sandstone, sandy pebble conglomerate with dispersed cobbles and boulders, and sporadic cobble conglomerates. Although predominantly matrix-supported, some thin framework conglomerates are also present and may show clast imbrication. Hubert (1966) records very rare cross-bedding and some reverse graded horizons are locally developed towards the base of the member. A comparable sequence to that deduced on the coast can be traced inland but it is everywhere poorly exposed. However, several fossiliferous localities are known and have been described in detail by Williams (1962).

A small outlier of blue-grey mudstone contained within the Stinchar Valley Fault zone west of Craigneil [NX 142 854] (Fig. 26) has been correlated with the Balclatchie Formation on the basis of its contained fauna (Peach and Horne, 1899; Williams 1962).

Ardwell Formation

The Ardwell Formation is impressively exposed along the full length of the shore between Kennedy's Pass and Ardwell Bay and is again seen on the northern side of the Bay at the base of the Whitehouse Formation. It attains a thickness of at least 750 m.

At Kennedy's Pass, the formation occupies a shallow channel with an irregular surface, excavated in the Kilranny Conglomerate, in which flaggy beds penetrate into small recesses overhung by the conglomerate (cf. Henderson, 1935). The lowest beds are dark green-grey siltstone and silty mudstone with numerous millimetre-thick, pale yellow, fine sandstone laminations. Less common are thin beds of medium-grained sandstone. Lenses of graded coarse sandstone and pebble conglomerate are present locally. Microfaults are numerous with sandstone commonly back-injected along the fault surfaces. The remainder of the sequence along the foreshore to the north-east is much sandier, formed of alternations 1 – 15 cm thick of medium-fine sandstone and siltstone with occasional black mudstone drapes. Normal grading is common with load and flame structures locally well developed. Parallel and cross-lamination are sufficiently common to suggest that, using turbidite terminology (Bouma, 1962), the sequence is characterised by T_{bce}, T_{ce} and, probably, T_{de} turbidites. Isolated coarser beds up to 75 cm thick are also present, typically as coarse sandstones with mudflake clasts, graded in their upper parts and showing complete T_{a-e} or base-absent T_{b-e} Bouma units. The turbulent passage of the current depositing some of these beds has shocked and fractured the underlying finer-grained rocks to a depth of 10 – 15 cm, with back-injection by the coarse sandy or gravelly matrix of the overlying unit (cf. Henderson, 1935, p.499). Graptolites and orthocones are widely distributed throughout the formation (Ingham, 1978) but are nowhere common.

Synsedimentary deformation structures are common and, in places, superbly displayed (Henderson, 1935). In addition to convolute bedding, lens-shaped intraformational breccias are locally well developed, comprising angular siltstone fragments and ovoid carbonate concretions dispersed in coarse sandy matrix. However, the most impressive structures are slump scars and associated collapse breccias that locally can be traced continuously for 60 m or more. The breccias form lensoid deposits up to 6 m thick, free of matrix and composed of large, jumbled and angular blocks of the flaggy Ardwell Formation beds. In places the breccias are cemented by large diagenetic carbonate nodules, attesting to

Figure 30 Sketch cross-sections through the Ardwell Formation exposed on the foreshore NE of Kennedy's Pass illustrating the style of folding, with no vertical exaggeration (*top*); together with detail of fold hinges (*bottom*) D 1563

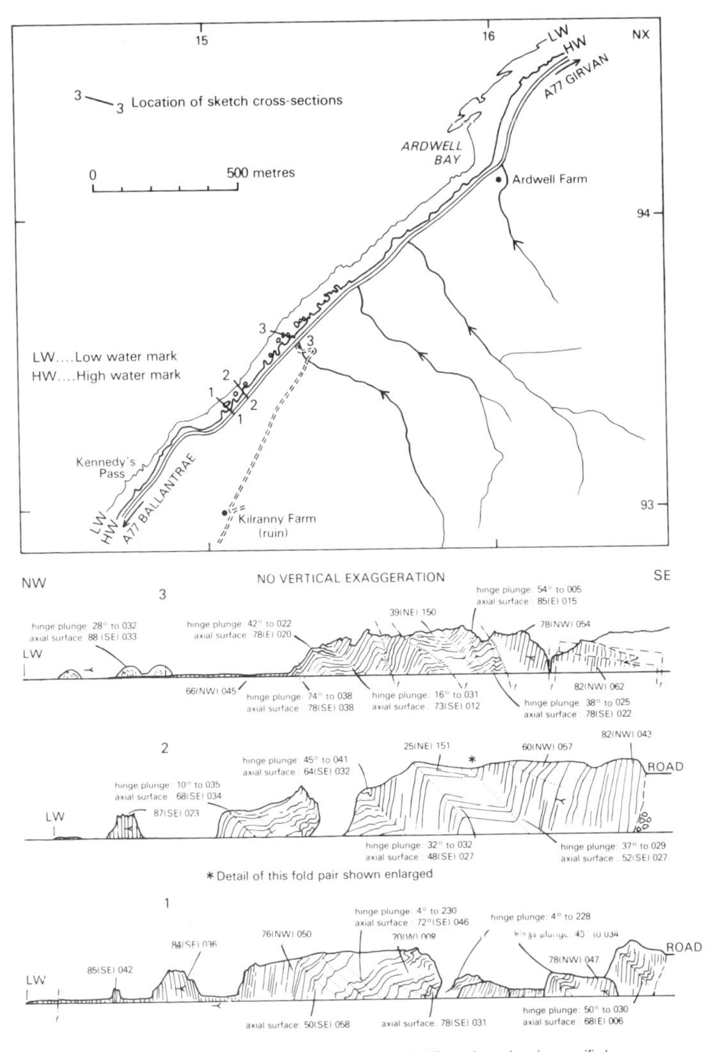

the synsedimentary origin of the deposits. The slump scars are sharply defined, with an arcuate, sometimes stepped, profile that has a shallow to moderately dipping attitude relative to original bedding orientations. Younger beds passively overlap the slump scars (Henderson, 1935, fig.15).

The Kennedy's Pass to Ardwell Bay shoreline contains 'the most impressive display of folding in the Girvan area' (Williams, 1959, p.631). There, concentric folding involving differential slip between the individual coarse and fine beds (flexural slip) has thrown the succession into a spectacular series of cascade or step folds with planar steep limbs and corrugated, shallow to moderately dipping limbs (Figure 30). The folds form a series of overlapping compressed periclines of no great lateral extent (tens of metres) that can be traced from inception to completion along the foreshore, except where interrupted by cross-faulting. The progressive rotation in fold axial plunges from shallow north-east, northwards to moderate or steep east-north-east, suggested two superimposed phases of deformation to A. Williams (1959): an early Ardwell 'main' fold phase characterised by the cascade folds, and a later phase responsible for the rotation of the fold axes. However, the structural analysis by G. D. Williams and Spray (1979), based on the critical assumption that the bedding-plane slickenside lineations were formed during the period of fold formation, demonstrated 'a straightforward deformational episode with no appreciable rotation of principal stress directions'. The deformation is thus non-cylindrical and a consequence of the folding of non-planar bedding surfaces, a situation that may have been created by soft sediment deformation prior to the cascade fold phase. An alternative view by Bluck (*in* Ingham, 1978) points to several enigmatic fold-related features, such as sand flow at fold closures, the emplacement of an axial planar sandstone dyke, and the stratigraphical confinement of the fold belt. From this evidence it was deduced that the sequence was deformed as a major, partly consolidated slump sheet.

Exposure of the Ardwell Formation inland is very poor, particularly in the large area between Knockbain, Pinmacher and High Letterpin. The rocks seen here are mainly siltstone and silty mudstone with numerous sandy laminations. These finer horizons are interbedded locally with thin, normally graded sandstones some of which are cross laminated. Interbedded conglomerates are known from a borehole near Knocklaugh Lodge (Stone and others, 1984) and are exposed in small quarries and natural exposures at High Letterpin.

Whitehouse Formation

Despite its small outcrop, confined to the foreshore north-east from Ardwell Bay, the Whitehouse Formation can be subdivided into three lithologically distinctive members that have been described in detail by Hubert (1966) and Ingham (1978). The lowest division, at least 65 m thick, consists of pale orange-brown, graded calcareous sandstone and interbedded pale grey-green shaly mudstone. Lower parts of sandstone beds may be pebbly with low-angle cross-stratification (Hubert, 1965, p.684) and the full association comprises a limestone flysch sequence. Complete T_{a-e} units are commonest in the thicker, sandy beds (30 cm – 1 m) with basal grading largely suppressed relative to the lower laminated division. The proportion of thinner-bedded, graded grey (terrigenous) sandstone and grey mudstone in T_{cde} and T_{de} divisions increases up the sequence, but prominent strike faulting cuts out an unknown thickness at the junction with the overlying sandstone member. The bedding is steeply dipping and slightly overturned, with numerous isolated fold pairs that seem to be related to minor dextral faulting. Larger areas of exceptionally disharmonic, open to tight folding with markedly varied fold plunges occur locally. Although affecting the topmost Ardwell Formation beds on the north side of Ardwell Bay, the folding seen in the Whitehouse Formation is less spectacular and dissimilar in style to the structures exposed south of Ardwell Bay. Identifiable shelly material is rare, but the trilobite *Tretaspis ceriodes*, of late Caradocian age, has been recovered from the calcareous sandstone (Ingham, 1978).

The succeeding, unfossiliferous sandstone member is also a flysch sequence, formed by about 30 m of thin but persistent, evenly spaced, grey sandstone and grey-green shaly mudstone. Most of the beds are T_{b-e}, T_{c-e} and T_{d-e} turbidites (Hubert, 1966). On the foreshore close to Ardmillan Lodge [NX 1680 9488], the thin-bedded sequence coarsens and thickens upwards into a unit 5 m thick of erosive-based, coarse gravelly sandstone and sandy pebble conglomerate with mudflake horizons and cross stratification. These beds, described here as the uppermost part of the middle sandstone member (cf. Hubert, 1966), were included as the basal beds of the overlying mudstone member by Ingham (1978).

The sandstone member passes up gradually through a transitional sequence of green shaly mudstone with impersistent thin sandstone horizons into an overlying member of green and red shaly mudstone with rare sandstone dykes. Sandstone laminations increase towards the top of the member, which is about 40 m thick. It contains a large indigenous fauna of graptolites, bizarre blind and large-eyed trilobites, and a restricted range of brachiopods (Ingham, 1978) indicating a late Caradoc to early Ashgill age.

Shalloch Formation

The Shalloch Formation (the Barren Flags of Lapworth, 1882, and Peach and Horne, 1899) is a flysch-like sequence of alternating fine to medium sandstone and pale green silty mudstone that transitionally overlies the Whitehouse Formation. Individual beds can be traced for a considerable distance (Ingham, 1978). The sandstones are classical turbidites, thicker-bedded and less spaced than those of the Whitehouse Formation (sandstone member), with superbly displayed internal structures corresponding mainly to T_{b-e} and T_{c-e} turbidites, with some sporadic T_{a-e} beds up to 1 m thick. Thin graded biosparite limestone horizons occur at a few places, one containing low Ashgill shelly debris. Rare graptolites (*Dicellograptus complanatus* and *D. anceps*; Ingham, 1978) occur locally. Only some 95 m of the sequence are exposed, the uppermost part being cut out by faulting.

The Tappins Group

South of the Stinchar Valley and Dove Cove Faults greywacke is the dominant rock type present, occurring in well bedded turbidite units (Bouma, 1962) up to about a metre in thickness. The lower parts of the units may be conglomeratic locally but more commonly consist of a medium- to coarse-grained greywacke, often with load casts at its base, grading upwards over several tens of centimetres to a finely laminated siltstone. This is overlain by a grey mudstone rarely more than a centimetre and usually only 2–3 mm thick. Cross lamination and convoluted bedding sometimes occur in the siltstones. The coarse greywacke base of the overlying turbidite unit abruptly succeeds the mudstone and may be erosive, leading in places to multiple or amalgamated beds. Peach and Horne (1899) first applied the term Tappins Group to the sequence cropping out between the Glen App and Dove Cove Faults. They considered it to be only about 150–200 m thick and attributed its extensive outcrop to multiple repetition by tight folding. However, a stratigraphic revision of the area by Williams (1962) disproved the fold model and established a succession several thousand metres thick.

Most of the greywackes are a dull greenish grey, but between the Stinchar Valley and the Dove Cove Faults red and green mudstone and dark purple pebbly greywacke alternate with green siltstone. This colour difference encouraged Williams (1962) to propose a subdivision of the Tappins Group into two different facies within the map area: the more colourful Traboyack division mainly between the Stinchar Valley and Dove Cove Faults; and the Dalreoch division cropping out between the Dove Cove and Glen App Faults (Figure 26). The validity and significance of these divisions are far from clear, however, the more so as Ince (1983) has used the terms in a slightly different sense and has proposed additional subdivisions. Apart from the local colour differences there is also some variation in the clast composition of the greywacke in Williams' two divisions. Both contain abundant quartz, feldspar and basaltic grains, but the Traboyack rocks in the north-west also commonly contain a varied assortment of coarse-grained igneous rock fragments, including serpentinite and granitic lithologies, whereas these are very much rarer in the Dalreoch greywackes.

The age evidence for the Tappins Group is very sparse. Williams (1962) considered that his Traboyack division was in part older than the Kirkland conglomerate whilst a further Tappins Group division, the Albany, recognised beyond the eastern margin of the map area and though to be younger than both the Traboyack and Dalreoch divisions, was equivalent to the upper part of the Stinchar Limestone Formation. On sedimentological grounds Ince (1983) thought the Traboyack division to be the lateral equivalent of the Kirkland Conglomerate and thus Llanvirn in age. No fossil evidence was obtained from the map area, but along strike to both north-east and south-west sparse graptolite faunas indicate a late Llandeilo to early Caradoc age. On this basis and on petrography the Tappins Group is broadly comparable to the northernmost units described elsewhere in the Southern Uplands: the Corsewall group of Kelling (1961) and the Marchburn Formation of Floyd (1982).

Despite local folding in the Tappins Group the steeply inclined beds generally strike NE–SW and young consistently towards the north-west. This overall structure is compatible with the inclusion of the group within an imbricate model for the Southern Uplands: either a forearc accretionary prism (Leggett and others, 1979) or a back-arc thrust duplex (Stone and others, in press). Either interpretation has the important implication that the Traboyack division, cropping out between the Stinchar Valley and Dove Cove faults, may be a separate thrust slice and the oldest part of the Tappins Group. The Tappins Group itself is likely to be amongst the oldest greywacke sequences in the Ordovician succession of the Southern Uplands.

The Leadhills Group

South east of the Glen App Fault the Leadhills Group flysch outcrop contains medium- and fine-grained greywacke, laminated siltstone and thin grey shale horizons, forming turbidite units up to about 50 cm thick. The greywackes contain more detrital quartz and fewer mafic rock or ferromagnesian mineral grains than the Tappins Group greywackes to the north. A very sparse graptolite fauna collected along strike to the south-west indicates an approximately basal Caradoc or uppermost Llandeilo age. No formal stratigraphy is proposed for this Leadhills Group greywacke turbidite sequence although is may well be equivalent to part of the Afton Formation described along strike to the north-east by Floyd (1982). The sequence cropping out in the map area is likely to be of the order of 1000 m thick.

Sedimentary sequences of Devonian and Permian age 7

Devonian

An unfossiliferous sedimentary sequence, traditionally assigned to the Lower Old Red Sandstone facies of early Devonian age, crops out in the north of the map area (Figure 26) in the vicinity of the farms at Glendrissaig [NX 191 949] and Pinminnoch [NX 188 939]. Interbedded mudstones, red sandstones, pebbly sandstones and conglomerates locally overlie unconformably the Llandeilo Benan Conglomerate Formation, but, for the most part, the Devonian sedimentary rocks are faulted against elements of the Ballantrae Complex or its upper Ordovician sedimentary cover. The sandstones are fine to medium grained, variably planar and cross-bedded. They are generally red but may in places be banded or mottled pale green or brown: a colour variation which is particularly marked in the finer-grained micaceous sandstones and their rare, silty mudstone interbeds. Discontinuous lenses of coarse sandstone in the sequence grade locally into pebbly sandstone and conglomerate containing rounded to subangular pebbles, cobbles and rare small boulders; all matrix-supported in a medium grained red sandstone. The clasts are a heterogeneous mixture of sandstone, mudstone, red granite, quartzite, chert, jasper, basalt, dolerite and greywacke.

Despite the truncation of the outcrop by faulting, the overall structure in the northern sector can be deduced as an open syncline plunging gently to the north. There is thus a general north-easterly dip in the west of the outcrop and a north-westerly dip in the east. Southwards, in the Pinminnoch fault block, no overall structure can be determined.

Permian

The south-western part of the Ballantrae Complex is faulted against and unconformably overlain by a sequence of red sandstones and breccias dipping uniformly to the north-east at about 15°. These strata have been assigned to the upper Permian or Lower Triassic (Smith and others, 1974; Brookfield, 1978) and are exposed in a narrow coastal strip (Figure 26) about 3 km long N–S by 0.5 km wide between Ballantrae Harbour [NX 082 830] and Bennane Lea [NX 862 859]. The outcrop forms the eastern fringe of a large Permo-Triassic basin underlying the North Channel and Firth of Clyde.

The lower part of the exposed succession is formed of the coarse clastic rocks of the Park End Breccia Formation, originally called the Ballantrae Breccia by Brookfield (1978) but here renamed to avoid nomenclature confusion with the Ballantrae Complex. The breccia unit is at least 250 m thick and consists of subrounded to subangular pebbles and cobbles, principally of greywacke, contained in a silt or fine sand matrix which also forms discontinuous but discrete horizons within the breccia. Accessory clast types include mudstone, chert, basalt and, more rarely, various granitic and gabbroic lithologies which tend to occur as small, well rounded cobbles. A crude imbrication of the clasts indicates deposition from a current flowing towards the north-west which Brookfield (1978) associated with a network of braided streams. Within the breccia a single horizon of dolerite (S 70491) exposed in Ballantrae Harbour [NX 0820 8305] may represent an original lava flow: if it does a slightly older age for the Formation is indicated than has previously been proposed since evidence for contemporaneous volcanicity is only seen elsewhere in south-west Scotland in the lower Permian (Mykura, 1965; Brookfield, 1978).

An exposure gap of approximately 200 m above the Park End Breccia Formation is succeeded by the Corseclays Sandstone Formation, at least 500 m thick and consisting of cross-laminated fine-grained red sandstone. The grains are mainly subrounded and of quartz or, more rarely, of plagioclase but a significant minority have a high degree of sphericity and are probably aeolian sand grains. These tend to be slightly larger than the average grain size, and they become slightly less rare to the north, i.e. up sequence.

The highest exposed beds of the Corseclays Sandstone Formation make up the Bennane Lea Breccia Member. This 10 m thick unit is distinctive in containing isolated angular to subrounded pebbles of basalt (S 71932, 71933) together with a few subrounded pebbles of greywacke and quartzite. All of the basaltic fragments are reddened and show a variety of textures. Some are almost entirely altered to chlorite, serpentine and clay minerals but preserve a relict doleritic texture. Serpentinite pebbles derived from an ultramafic protolith were reported by Peach and Horne (1899).

Brookfield (1978) considered the Corseclays Sandstone Formation to represent channel and overbank deposits of predominantly sandbed, ephemeral streams. There was clearly a contribution of wind-blown grains but the larger and more angular basaltic clasts in the Bennane Lea Member were probably locally derived. The lithologies are not sufficiently distinctive to allow the identification of their provenance but the serpentinite clasts described by Peach and Horne (1899) seem likely to have been derived from the neighbouring Ballantrae Complex. The basalt clasts could readily have been derived from the same source.

The southern and eastern margins of the Permian outcrop are now in faulted contact with the Ballantrae Complex, but the northern contact, exposed below high water mark at Bennane Lea, is more ambiguous. Some faulting has taken place along the contact but the detailed relationship is more suggestive of an unconformity. The overall dip of the Permian sandstone and breccia beds (about 15° towards the north-east) militates geometrically against a simple unconformable relationship, and an element of northerly transgression of the Permian strata across the Ballantrae Complex may be present.

Post-Ordovician minor intrusions 8

Devonian

A small number of highly altered porphyritic and felsic dykes, 0.5 to 2 m thick, are intruded into both the Ballantrae Complex and the Lower Palaeozoic greywacke sequence to the south of the Stinchar Valley Fault. These dykes are considered to be members of a regional 'Caledonian' dyke swarm emplaced in earliest Devonian times (Richey, 1939). South of the Stinchar Valley Fault the dykes contain a variable quantity of sericitised plagioclase phenocrysts and chlorite pseudomorphs after original mafic phenocrysts, all enclosed in a matrix which is itself extensively chloritised and carbonatised. Richey (1939) described such rocks as porphyrites within a lithologically mixed, NE-trending dyke swarm cropping out widely in south-west Scotland. However, the examples seen in the southern part of the map area have a more variable orientation ranging from north-east to north-west. Farther north, within the Ballantrae Complex a group of felsic dykes crops out near Balsalloch [NX 123 884] with an approximate north-south trend. In these rocks the proportion of phenocrysts is much lower and those present are exclusively of altered plagioclase, whilst quartz is an important constituent of the matrix. Similar dykes are again described by Richey as common members of the Caledonian swarm of south-western Scotland.

Permo-Carboniferous

A single alkaline lamprophyre dyke, trending N–S and 3 m thick, crops out [NX 1464 9267] south-west of Kennedy's Pass (specimen and unpublished data provided by Mr A. Herriot) as an intrusion into lavas and breccias of the Balcreuchan Group. In thin section (Ed 7235) cumulophyric aggregates of clinopyroxene and greenish brown amphibole are contained in a fine-grained groundmass dominated by plagioclase, nepheline, analcime and apatite. Both nepheline and apatite also occur as small, isolated euhedral phenocrysts. The dyke seems likely to be an outlier of the large Permo-Carboniferous swarm of similar lithology seen extensively in the western Highlands and parts of central Scotland.

Tertiary

The youngest rocks intruding the Ballantrae Complex are a group of broadly NW-trending basaltic dykes of probable Tertiary age. The dykes are intruded into all elements of the Complex, the greywacke sequences to the south and the Permian succession of the Ballantrae area. Tholeiites and olivine dolerites are both represented, the tholeiites being slightly the more common, but there is no significant difference in their geographical distribution or orientation. Thus, a bimodal division between north-west and north-north-west within the main trend is not related to compositon. Similarly, where intersecting dykes can be seen on the foreshore north of Ballantrae [NX 0845 8335], both sets are tholeiitic, although those with an approximate north-east trend appear to be earlier than those trending generally north-west. The orientation of several dykes is clearly controlled by pre-existing structures, since they intrude the fault zones separating the major units of the Ballantrae Complex. A good example of this is to be seen on the shore at Pinbain Bridge [NX 1370 9137], where tholeiitic dykes intrude the tectonic contacts between serpentinite and an Arenig mélange sequence of the Balcreuchan Group.

In thickness the dykes range up to about 2 m, but the majority are less than 1 m and many are only a few tens of centimetres across. They are variably vesicular and in thin section show a mass of relatively fresh plagioclase, augite, ore minerals and more rarely olivine, enclosing microphenocrysts of plagioclase and occasionally olivine. Flow banding is extremely well developed in some specimens but entirely absent in most. All of the dykes are probably associated with the large Arran and Mull swarms (Richey and others, 1961), members of which have a wide distribution in south-west Scotland.

The evolution of the Ballantrae Complex: summary and discussion

9

Age control

The maximum ages so far reported from the Ballantrae Complex are the 576 ± 32 Ma and 505 ± 11 Ma Sm-Nd dates obtained by Hamilton and others (1984) from garnet-clinopyroxenite segregations in harzburgite at Knockormal and Knocklaugh respectively (Bloxam and Allen, 1960; Smellie and Stone, 1984). This suggests the inclusion within the Complex of some mantle material generated in the Cambrian, but the bulk of age evidence, both radiometric and biostratigraphical, shows that most of the elements forming the Complex originated during the early Ordovician. This evidence is summarized in Figure 31 together with time ranges for the Arenig calculated by Harland and others, (1982) and McKerrow and others (1985).

The oldest eruptive rocks present are likely to be the mature island-arc basalts of the Mains Hill Agglomerate Formation (Balcreuchan Group) on the evidence of the 501 ± 12 Ma Sm-Nd age obtained by Thirlwall and Bluck (1984). These may very well overlap in age with the oceanic island-type basalts of the Bennane Head and Pinbain areas (Stone and Rushton, 1983; Rushton and others, 1986), which have been palaeontologically dated as early to middle Arenig. It is also difficult to separate the age of the 476 ± 14 Ma Games Loup primitive island-arc basalt sequence (Thirlwall and Bluck, 1984) and the 478 ± 8 Ma date from an amphibolite in the metamorphic aureole of the northern serpentinite outcrop, which has been taken as the time of obduction of the Complex (Bluck and others, 1980). The late Arenig graptolite fauna and its containing sedimentary sequence at North Ballaird may by either younger (McKerrow time-scale) or older (Harland time-scale) than the supposed obduction event (Figure 31) and so may or may not correlate with the Games Loup succession. Algal coatings on rounded serpentinite pebbles (Figure 32) in interbedded conglomerate from the North Ballaird sequence certainly imply that serpentinite was exposed subaerially or in shallow water by the late Arenig and so favour McKerrow. This situation is compatible with the 483 ± 4 Ma date obtained from trondhjemite (Bluck and others, 1980) intrusive into the ultramafic body to which the metamorphic aureole is marginal. The trondhjemite is transitional through quartz diorite to a marginal gabbro which is chilled against the

Figure 31 Summary diagram of age data from the Ballantrae Complex

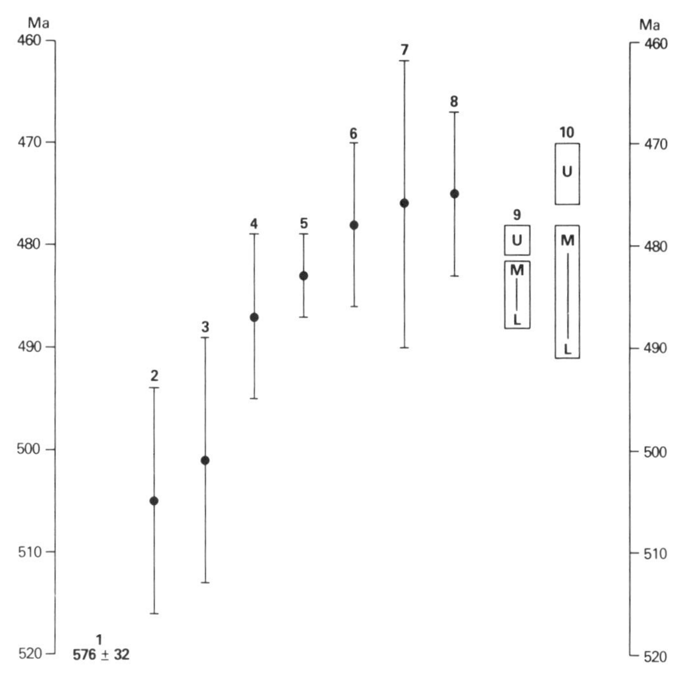

1 . . . . . Sm – Nd, Garnet-clinopyroxenite. Hamilton and others, 1984
2 . . . . . Sm – Nd, Metapyroxenite. Hamilton and others, 1984
3 . . . . . Sm – Nd, Basalt. Thirlwall and Bluck, 1984
4 . . . . . K – Ar, Flaser gabbro. Bluck and others, 1980
5 . . . . . U – Pb, Trondhjemite. Bluck and others, 1980
6 . . . . . K – Ar, Amphibolite. Bluck and others, 1980
7 . . . . . Sm – Nd, Basalt. Thirlwall and Bluck, 1984
8 . . . . . K – Ar, Gabbro. Harris and others, 1965
9 and **10**........Fossiliferous sequences of Lower to Middle and Upper Arenig age (Stone and Rushton, 1983) placed in the Arenig time scales proposed by Harland and others, 1982 **(9)** and by McKerrow and others, 1985 **(10)**

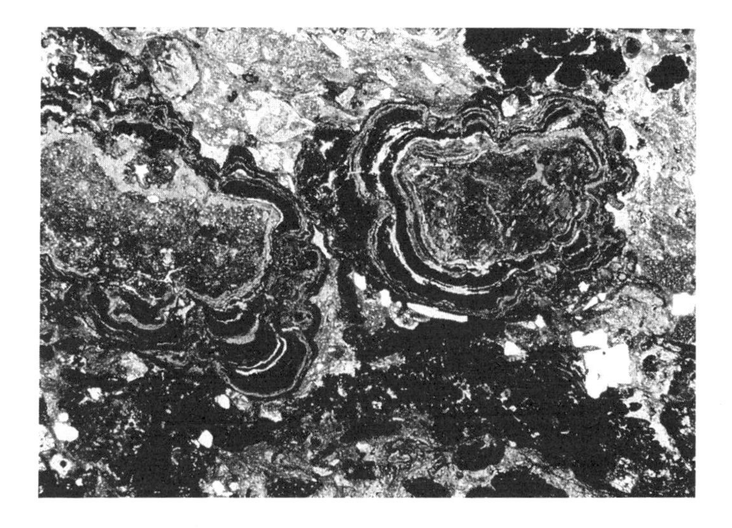

Figure 32 Algal coatings on serpentinite pebbles from the late Arenig sequence at North Ballaird. Plane polarised light, ×20 S 67687, PMS 295

serpentinite, thus requiring the serpentinite to have been cool at 483 ± 4 Ma. However, the ultramafic rock is required to have been hot at 478 ± 8 Ma to produce the metamorphic aureole. Error overlap allows both requirements to be met, but the onset of thrusting and the formation of the aureole is therefore likely to have occurred before 480 Ma.

Enigmatic mélange deposits containing blueschist fragments of unknown provenance are now in tectonic contact with the margins of the northern serpentinite outcrop; they also contain amphibolite clasts believed to have been derived from the 478 ± 8 Ma metamorphic aureole. The mélange deposits can therefore be no older than middle to late Arenig whichever time scale is preferred, although the faunal control (Rushton and others, 1986) is inadequate to justify the use of this association as a constraint on the age of the Arenig as suggested by Bluck and others (1980). A middle or late Arenig age is similarly implied for the Craig Hill Breccia Formation (and the basalt lavas with which it is interbedded) if the granitic clasts which it contains have indeed been derived from the 483 ± 4 Ma Byne Hill trondhjemite body.

That the 478 ± 8 Ma aureole was not formed during the final emplacement (?obduction) of the Ballantrae Complex is shown by the suite of dykes which intrude the northern serpentinite and cut the metamorphic aureole at its base (Chapter 4, see the first section). These have been geochemically divided into two groups (Holub and others, 1984), the larger having characteristics of 'within-plate' oceanic-island basalts and the smaller being more similar to primitive island-arc basalts. Either of these intrusive suites could be associated with the late Arenig tuffaceous sequence from North Ballaird, and the second (smaller) may correlate with the Games Loup Lavas. However, it is difficult to reconcile the formation of tuffs and presumably oceanic-island and island-arc basalts above broadly 'oceanic' lithosphere already obducted onto a continental margin. The 478 Ma date is therefore likely to relate only to the initiation of thrusting within the oceanic crust (cf. Spray and Williams, 1980); final emplacement of the Complex must have been a later event. This is constrained by the Llanvirn age of the unconformably overlying Kirkland Conglomerate Formation and so fits most readily as a late Arenig event on the time scale of McKerrow and others (1985). There are considerable problems in squeezing all of the known data points into the late Arenig section of the alternative Harland and others (1982) scheme.

Petrological and textural differences between the two main ultramafic outcrops (the northern and southern serpentinite bodies, described in Chapter 2) are compatible with their formation at markedly different structural levels within the ultramafic zone of oceanic crust generated at a spreading ridge: the northern serpentinite representing a much deeper level than the southern belt. They could therefore be parts of a classical ophiolite sequence, a possibility strengthened by the intrusion of trondhjemite into the northern ultramafic outcrop at about 483 Ma (Bluck and others, 1980) and favoured on geochemical grounds by Jelinek and others (1984). However, xenoliths, possibly fragments of a thermal aureole, provide evidence from the southern outcrop for the possible intrusion of hot, mobile peridotite into the uppermost part of a pre-existing ultramafic cumulate sequence and its overlying transition zone to gabbro (see Chapter 2, Basic igneous xenoliths . . .). Such an event may be dated by the 487 ± 8 Ma age of metamorphosed flaser gabbro (Bluck and others, 1980), which is found as a very large xenolith in the serpentinite at Millenderdale [NX 175 905], and the 475 ± 8 Ma age from a poorly located and described 'gabbro' collected north of Colmonell by Harris and others (1965). The Cr:Al ratios from the ubiquitous accessory spinel minerals (Chapter 2, Economic mineralisation . . .) support such a polygenetic origin for the southern serpentinite outcrop (Dick and Bullen, 1984; Stone and others, 1986), whereas the Cr:Al data from spinels in the northern outcrop generally favour the generation of that ultramafic rock in a sub-arc environment rather than beneath an oceanic spreading ridge. Analagous situations for more recent 'island arc ophiolites' have been reported from the Marianas trench (Bloomer and Hawkins, 1983) and the Philippine Islands (Hawkins and Evans, 1983). Thus the differences in the two main ultramafic outcrops may relate to origins in different petrotectonic settings rather than different structural levels in the same setting. From either interpretation it is clear that the two main ultramafic bodies in the Ballantrae Complex were originally spatially separated and are unlikely to be parts of a once laterally continuous unit.

Structural development

The Ballantrae Complex contains a number of disparate elements generated in different petrotectonic environments at approximately the same time. Conversely there is also evidence for the inclusion of sequences from the same geological setting but of different ages. This has been well established for the volcano-sedimentary Balcreuchan Group but is probably equally true for the ultramafic parts of the Complex. There is evidence for thrusting during the Arenig, and this has been related to the imbrication of at least part of the Balcreuchan Group (Stone and Rushton, 1983). Some of the middle or upper Arenig clastic sedimentary sequences interbedded with the extrusive lavas contain clasts presumed to have a relatively local origin, e.g. the amphibolites in the Pinbain mélange and the granitic rocks in the Craig Hill Breccia Formation, suggesting that the site of deposition of the clastic rocks was close to the partially assembled, older parts of the Ballantrae Complex. However, those older elements were themselves brought together from originally widely spaced sites, and several models have been proposed to explain their present proximity. A marginal-basin volcanic-arc association has been suggested by Bluck and others (1980), the basin and arc being formed and then thrust onto a continental margin between 490 and 470 Ma. This would be compatible with the generation of much of the ultramafic rock in a sub-island-arc environment. As alternative explanations Barrett and others (1982) prefer the impingement of intraplate volcanic edifices on a subduction trench, and Stone (1984) has argued in favour of substantial strike-slip movement, pointing to the alternation of wrench and thrust tectonism in large-scale, strike-slip terrains such as the Alpine fault of New Zealand (Carter and Norris, 1976) or the Turkish Antalya complex (Woodcock and Robertson, 1982).

Whatever the mechanism, the entire Ballantrae Complex had been assembled and emplaced at a continental margin as a series of imbricate thrust sheets, by the late Arenig or earliest Llanvirn. A complicated structural history during thrusting is evinced by the four deformation episodes described from the metamorphic aureole (Spray and Williams, 1980), effectively a dynamothermal mylonite zone. Some idea of the gross structure at that time can be gained by a consideration of the geometrical relationships at the unconformable contact of the Barr Group and the Ballantrae Complex between Aldons Hill [NX 190 910] and Knockbain [NX 162 900]. If the unconformity is stereographically restored to the horizontal the contact between the Balcreuchan Group and the southern ultramafic outcrop rotates to dip at about 20° to the west. Above this unconformity the Barr Group was deposited in an extensional half-graben with fault-controlled transgression to the north-west (Williams, 1962; Ingham, 1978) across the structurally inclined Ballantrae Complex. Sedimentation was partly controlled by the contemporaneous faulting, and the whole succession was then affected by Caledonian deformation, which has been divided by A. Williams (1959) into two fold episodes followed by two wrench-fault episodes, all followed by three phases of normal faulting. Much of the evidence was derived from outcrop beyond the eastern margin of the map area considered here, but the nature of the second of Williams' fold episodes was largely deduced from the coastal Ardwell Formation outcrop (Figure 30). Subsequently G. D. Williams and Spray (1979) showed that all of the structures there could be accommodated within a single deformation phase. North-west-directed thrusting has affected the Barr and Ardmillan groups and may be expected to have had some influence on the underlying Ballantrae Complex. This event was thought by A. Williams (1959) to be the culmination of the major fold phase, but the thrust planes are themselves folded, and their outcrop pattern suggests a similar geometry to that of the folds affecting the basal Barr Group unconformity. An alternative explanation may therefore be an early thrusting event on low-angle planes with movement directed to the north-west, followed by a major folding episode on NE–SW axes. It may have been during this 'Caledonian' folding that the elements of the Ballantrae Complex were brought to their present apparently steep attitudes. That the thrusting and folding were essentially shallow, thin-skinned processes is established by the unusually low values of the colour alteration index reported for conodonts in the Barr and Ardmillan Group sediments by Bergström (1980). These show that the sedimentary rocks have at no time been heated above 100°C, equivalent to a maximum burial depth of perhaps 3000 m.

An additional constraint on the present disposition of the ultramafic bodies at depth is provided by their geophysical responses, notable gravity (Figure 33) and aeromagnetic (Figure 34). The aeromagnetic data for the northern serpentinite mass show a marked positive anomaly, and there is a negative anomaly with respect to the regional gravity. From these the northern serpentinite mass may be deduced to have a substantial continuation at depth. The southern outcrop however, is poorly defined magnetically and coincides approximately with a gravity high; only a slight disturbance on the northern margin of the positive anomaly (Figure 33) coincides with a part of the ultramafic outcrop. Thus the southern serpentinite may be a fairly shallow body with no significant deep continuation, a possibility first outlined by Powell (1978).

Faulting (probably post-Ordovician) has further complicated the structure of the southern serpentinite outcrop by introducing slices of the Balcreuchan and Barr Group successions in a linear zone from Garnaburn [NX 155 869] in the north-east to Bougang [NX 115 855] in the south-west. This faulting was probably associated with late movement on the adjacent and parallel Stinchar Valley Fault, a structure which also contains tectonic slices of both serpentinite and the younger, upper Ordovician sedimentary rocks. Other faults trending approximately north–south, themselves regarded as late structures within the Ballantrae Complex (Williams, 1959), are truncated by the Stinchar Valley Fault, which may also be the site of substantial Silurian

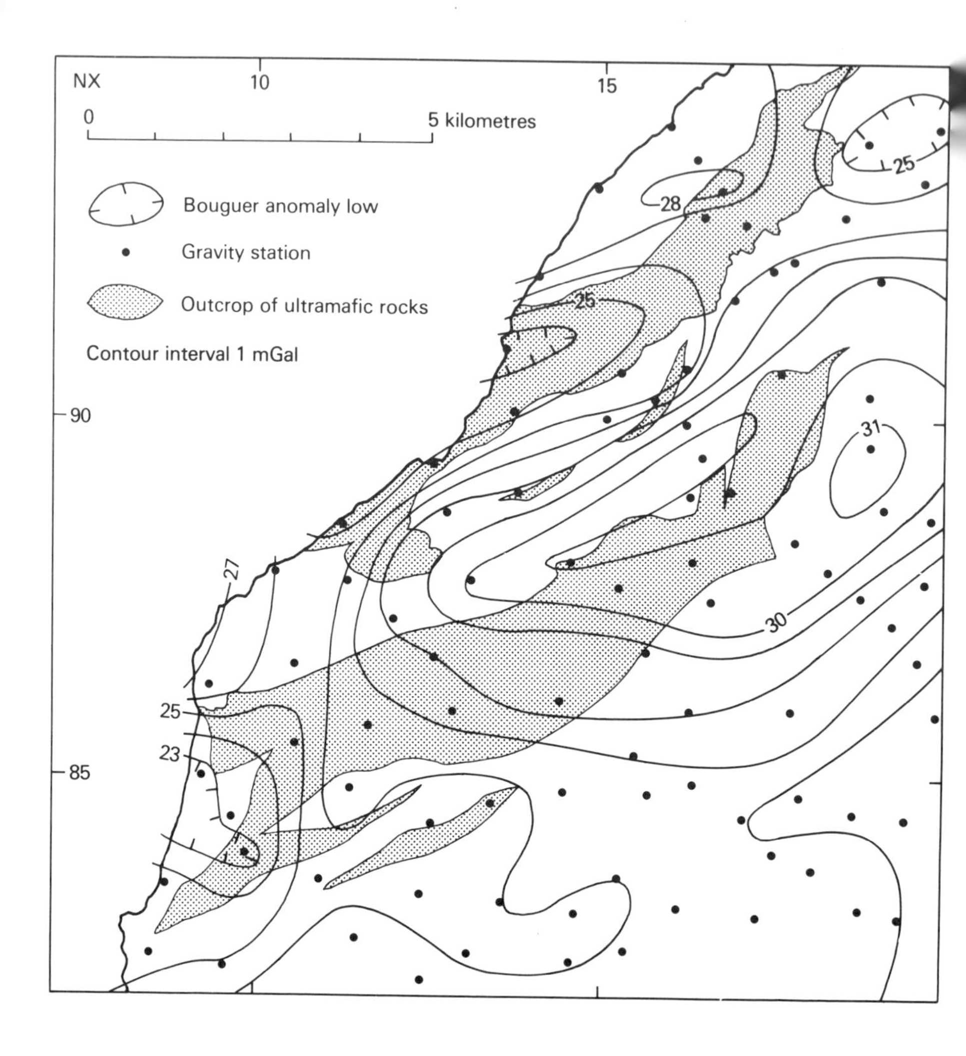

strike-slip movement (Leggett and others, 1983; cf. Dewey and Shackleton, 1984).

Figure 33 Bouguer gravity anomaly map of the Ballantrae area (cf. British Geological Survey, 1985)

Despite the apparent plethora of structural information from the Ballantrae Complex it is still not possible to predict the subsurface disposition of even the main lithological units with any degree of confidence. Early attempts by Peach and Horne (1899) and Bailey and McCallien (1957) are most certainly over-simplifications, and although the multiple-thrust models of Dewey (1974), Jones (1977) and Bluck and others (1980) are more realistic, they too probably fail to recognise the true level of complication within the Complex. However, a likely sequence of events for its genesis and assembly can be deduced and is summarised in Table 6.

The southern margin of the Ballantrae Complex is marked by the Stinchar Valley Fault, a major lineament which

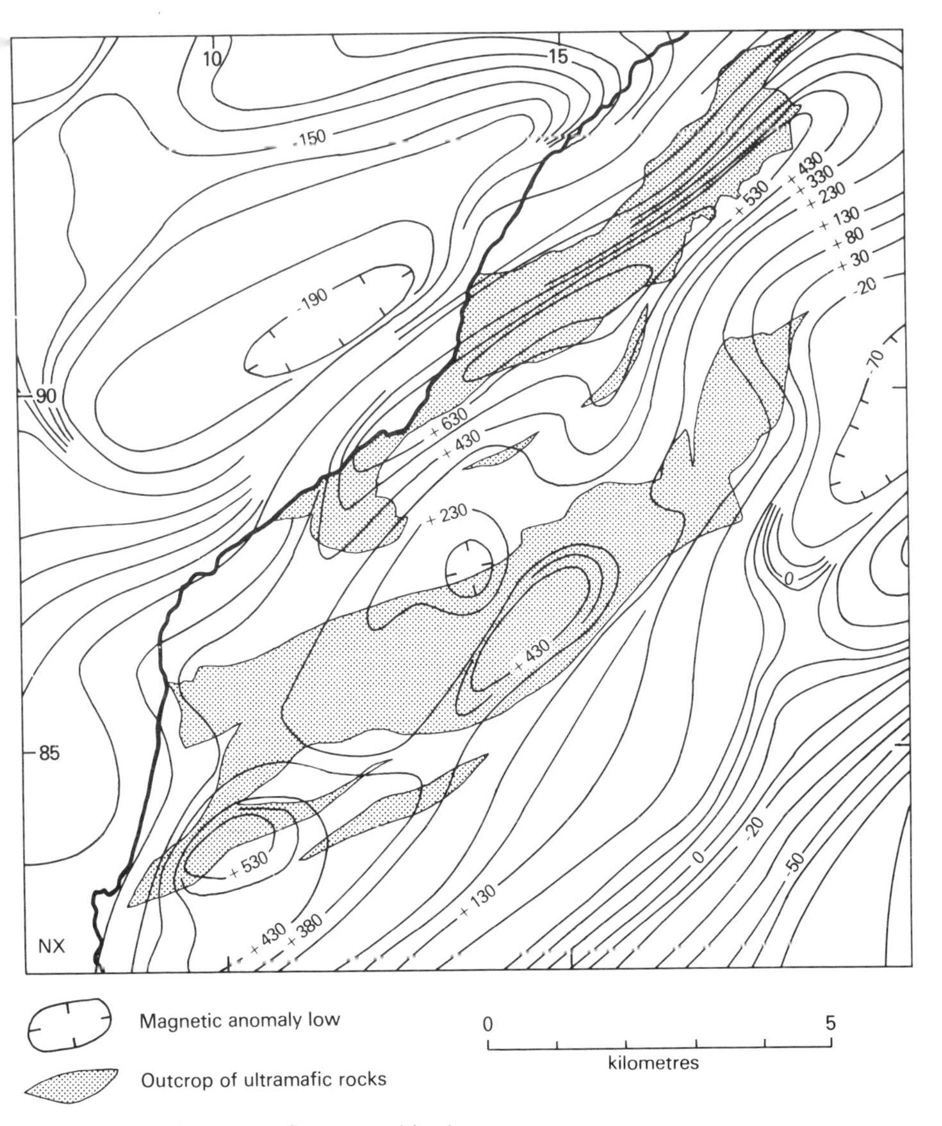

Figure 34 Regional aeromagnetic anomaly map of the Ballantrae area (cf. Institute of Geological Sciences, 1980)

divides contrasting geological terrains: to the north a zone dominated by Arenig and later thrusting directed to the north-west (the Ballantrae Complex together with the Barr and Ardmillan groups), and to the south a zone dominated by post-Llandeilo thrusting directed to the south-east (the Southern Uplands imbricate terrain).

Additional post-emplacement structural complexity can be deduced from the presence above the Ballantrae Complex of folded lower Old Red Sandstone strata, probably evidence of

Table 6 A summary of the principal events in the formation of the Ballantrae Complex. This is a composite tabulation and does not represent a continuous sequence at any one place. For full details of the age control see Figure 31

Event	Age Control
13 Thrusting and folding in response to 'Caledonian' NW–SE compression	Siluro-Devonian
12 Deposition of sedimentary sequences unconformably above the Ballantrae Complex	Llanvirn onwards
11 Alternating thrust imbrication and strike-slip faulting during the final assembly of the Complex. ?Serpentinisation	
10 Deposition of tuffs, breccias and mélange deposits in part derived from the embryonic Ballantrae Complex	Upper Arenig
9 Serpentinisation of the ultramafic rocks and associated Ca-metasomatism	
8 Eruption of primitive island-arc basalts (Games Loup)	476 ± 14 Ma
7 Possible structural imbrication	
6 Intrusion of trondhjemite, gabbro and minor doleritic dykes into the cooled ultramafic rock of the northern outcrop	483 ± 4 Ma
5 Initiation of NW-directed thrusting of hot ultramafic rock within oceanic crust. Imbrication and/or strike-slip movement in continental margin transform zone	478 ± 8 Ma
4 Possible intrusion of hot and mobile peridotite into the transition zone, gabbros and dyke complex of an oceanic ophiolite, possibly in a sub-arc setting	475 ± 8 Ma 487 ± 8 Ma
3 Eruption of lavas in an oceanic island, Hawaiian-type setting. Extensive penecontemporaneous brecciation. (Bennane Head and Pinbain)	Lower–Middle Arenig
2 Eruption of mature island-arc basalts (Mains Hill Agglomerate Formation)	501 ± 12 Ma
1 Crystallisation of ultramafic mantle segregations. Probably the initiation of cooling in the oceanic crust and mantle fragment now preserved as ophiolite units within the Ballantrae Complex	505 ± 11 Ma

mid-Devonian deformation, and from the inclined attitude of the Permian strata overlying the western margin of the southern ultramafic outcrop. This inclination may simply be the result of roll-over during Permian basin development and associated contemporaneous faulting, but it could also represent a regional tilting which would, of course, have affected the attitude of all of the features within the Ballantrae Complex.

References

ANDERSON, J. G. C. 1936. Age of the Girvan-Ballantrae serpentine. *Geol. Mag.*, Vol.73, 535–545.

ANDERTON, R., BRIDGES, P. H., LEEDER, M. R., and SELLWOOD, B. W. 1979. *A dynamic stratigraphy of the British Isles. A study in crustal evolution.* (London: George Allen and Unwin.)

BAILEY, E. B., and McCALLIEN, W. J. 1952. Ballantrae problems: historical review. *Trans. Edinburgh Geol. Soc.*, Vol.15, 14–38.

BAILEY, E. B., and McCALLIEN, W. J. 1954. External metasomatism associated with serpentine. *Nature, London*, Vol.174, 836.

BAILEY, E. B., and McCALLIEN, W. J. 1957. The Ballantrae serpentine, Ayrshire. *Trans. Edinburgh Geol. Soc.*, Vol.17, 33–53.

BAILEY, E. B., and McCALLIEN, W. J. 1960. Some aspects of the Steinmann Trinity, mainly chemical. *Q. J. Geol. Soc. London*, Vol.124, 365–395.

BALSILLIE, D. 1932. The Ballantrae igneous complex, south Ayrshire. *Geol. Mag.* Vol.69, 107–131.

BALSILLIE, D. 1937. Further observations on the Ballantrae igneous complex, south Ayrshire. *Geol. Mag.*, Vol.74, 20–33.

BARNES, I., and O'NEIL, J. R. 1969. The relationship between fluids in some fresh alpine-type ultramafics and possible modern serpentinization, western United States. *Bull. Geol. Soc. Am.*, Vol.80, 1947–1960

BARNES, I., RAPP, J. B., and O'NEIL, J. R. 1972. Metamorphic assemblages and the direction of flow of metamorphic fluids in four instances of serpentinization. *Contrib. Mineral. Petrol.*, Vol.35, 263–276.

BARRETT, T. J., JENKYNS, H. C., LEGGETT, J. K., and ROBERTSON, A. H. F. 1981. Volcanic stratigraphy and possible tectonic setting of the Ballantrae ophiolitic complex, S. W. Scotland. *Ofioliti*, Vol.6., 165–166.

BARRETT, T. J., JENKYNS, H. C., LEGGETT, J. K., and ROBERTSON, A. H. F. 1982. Comment on 'Age and origin of Ballantrae ophiolite and its significance to the Caledonian orogeny and the Ordovician time scale'. *Geology*, Vol.10, 331.

BASU, A. R. 1977. Textures, microstructures and deformation of ultramafic xenoliths from San Quintin, Baja California. *Tectonophysics*, Vol.43, 213–246.

BERGSTRÖM, S. M. 1971. Conodont biostratigraphy of the Middle and Upper Ordovician of Europe and eastern North America. *Mem. Geol. Soc. Am*, No.127, 83–157.

Bergström, S. M. 1980. Conodonts as palaeotemperature tools in Ordovician rocks of the Caledonides and adjacent areas in Scandinavia and the British Isles. *Geol. Foren. Stockholm Forh.*, Vol.102, 377–392.

Beveridge, R. 1950. On the igneous rocks of Arenig age in the Girvan-Ballantrae district, Ayrshire, and their stratigraphical relationships. Unpublished PhD thesis, University of Edinburgh, 185 pp.

Bloomer, S. H., and Hawkins, J. W. 1983. Gabbroic and ultramafic rocks from the Mariana Trench: an island arc ophiolite. 294–317 *in* The tectonic and geologic evolution of southeast Asian seas and islands: Part 2. Hayes, D. E. (editor). *Geophys. Monogr. Am. Geophys. Union*, No.27.

Bloxam, T. W. 1954. Rodingite from the Girvan-Ballantrae Complex, Ayrshire. *Mineral. Mag.*, Vol.30, 525–528.

Bloxam, T. W. 1955. The origin of the Girvan-Ballantrae beerbachites. *Geol. Mag.*, Vol.92, 329–337.

Bloxam, T. W. 1960. Pillow-structure in spilitic lavas at Downan Point, Ballantrae. *Trans. Geol. Soc. Glasgow*, Vol.24, 19–26.

Bloxam, T. W. 1964. Hydrogrossular from the Girvan-Ballantrae complex, Ayrshire. *Mineral. Mag.*, Vol.33, 814–815.

Bloxam, T. W. 1968. Petrology of Byne Hill, Ayrshire. *Trans. R. Soc. Edinburgh*, Vol.68, 105–122.

Bloxam, T. W. 1980. Amphibolite contact zones, amphibolite xenoliths, and blueschists associated with serpentinite in the Girvan-Ballantrae complex, southwest Scotland. *Arch. Sci.*, Vol.33, 291–299.

Bloxam, T. W., and Allen, J. B. 1960. Glaucophane schist, eclogite and associated rocks from Knockormal in the Girvan-Ballantrae complex, south Ayrshire. *Trans. R. Soc. Edinburgh,* Vol.64, 1–27.

Bluck, B. J. 1978. Geology of a continental margin 1: the Ballantrae Complex. 151–162 in *Crustal evolution in northwestern Britain and adjacent regions.* Bowes, D. R., and Leake, B. E. (editors). *Geol. J. Spec. Issue*, No.10.

Bluck, B. J. 1982. Hyalotuff deltaic deposits in the Ballantrae ophiolite of SW Scotland: evidence for crustal position of the lava sequence. *Trans. R. Soc. Edinburgh: Earth Sci.* Vol.72 (for 1981), 217–228.

Bluck, B. J. 1983. Role of the Midland Valley of Scotland in the Caledonian orogeny. *Trans. R. Soc. Edinburgh: Earth Sci.*, Vol.74, 119–136.

Bluck, B. J., Halliday, A. N., Aftalion, M., and Macintyre, R. M. 1980. Age and origin of Ballantrae ophiolite and its significance to the Caledonian orogeny and Ordovician time scale. *Geology,* Vol.8, 492–495.

Bonney, T. G. 1878. On the serpentine and associated igneous rocks of the Ayrshire coast. *Q. J. Geol. Soc. London*, Vol.34, 769–785.

Bouma, A. H. 1962. *Sedimentology of some flysch deposits; a graphic approach to interpretation.* 168pp. (Amsterdam: Elsevier.)

British Geological Survey. 1985. 1:250 000 Bouguer gravity anomaly map (provisional edition), Clyde sheet, 55°N – 06°W (Southampton: Ordnance Survey.)

Brookfield, M. E. 1978. Revision of the stratigraphy of Permian and supposed Permian rocks of southern Scotland. *Geol. Rundsch.*, Vol.67, 110 – 149.

Carruthers, R. M. 1980. Detailed airborne surveys, 1978. An assessment of geophysical data from the Girvan-Ballantrae district. *Rep. Appl. Geophys. Unit. Inst. Geol. Sci.*, No.78, (Unpublished.)

Carter, R. M., and Norris, R. J. 1976. The Cainozoic history of southern New Zealand; an accord between geological observation and plate tectonic prediction. *Earth & Plant. Sci. Lett.*, Vol.31, 85 – 94.

Church, W. R., and Gayer, R. A. 1973. The Ballantrae ophiolite. *Geol. Mag.*, Vol.110, 497 – 510.

Coleman, R. G. 1967. Low-temperature reaction zones and alpine ultramafic rocks of California, Oregon and Washington. *Bull. US Geol. Surv.*, No.1247, 1 – 49.

Coleman, R. G. 1977. *Ophiolites: ancient oceanic lithosphere?* 229pp. (Berlin, Heidelberg and New York: Springer-Verlag.)

Coleman, R. G., and Keith, T. E. 1971. A chemical study of serpentinization—Burro Mountain, California. *J. Petrol.*, Vol.12, 311 – 328.

Cooper, R. A., and Fortey, R. A. 1982. The Ordovician graptolites of Spitsbergen. *Bull. Br. Mus. (Nat. Hist.) (Geol.)*, Vol.36, 157 – 302.

Cortesogno, L., Galbiati, B., and Principi, G. 1981. Detailed description of some characteristic outcrops of Eastern Liguria serpentinitic breccias and geodynamic interpretation. *Ofioliti*, Vol.6, 47 – 76.

Dewey, J. F. 1974. The geology of the southern termination of the Caledonides. 205 – 231 in *The ocean basins and their margins, Vol.2: The north Atlantic.* Nairn, A. (editor). (New York and London: Plenum Press.)

Dewey, J. F., and Shackleton, R. M. 1984. A model for the evolution of the Grampian tract in the early Caledonides and Appalachians. *Nature, London,* Vol.312, 115 – 121.

Dick, H. J. B., and Bullen, T. 1984. Chromian spinel as a petrogenetic indicator in abyssal and alpine-type peridotites and spatially associated lavas. *Contrib. Mineral. Petrol.* Vol.86, 54 – 76.

Dixon, J. 1980. Aspects of metamorphism associated with the Ballantrae Complex. Unpublished MSc thesis, University of Wales. 199 pp.

Fettes, D. J. 1978. Caledonian and post-Caledonian metamorphism in the United Kingdom and Ireland. 75 – 85 in *Metamorphic map of Europe (1:2 500 000): explanatory text.* Zwart,

H. J., SOBOLEV, V. S., and NIGGLI, E. (editors). (Leiden: CGMW subcommission for the cartography of the metamorphic belts of the world; *and* Paris: Unesco.)

FLOYD, J. D. 1982. Stratigraphy of a flysch succession: the Ordovician of W Nithsdale, SW Scotland. *Trans. R. Soc. Edinburgh: Earth Sci.*, Vol.73, 1–9.

GEIKIE, A. 1897. *Ancient volcanoes of Great Britain*, Vol.1. 477pp. (London: McMillan.)

GEIKIE, A., and GEIKIE, J. 1869. Explanation of Sheet 7. Ayrshire: south-western district. *Mem. Geol. Surv. Scotland*, 16 pp.

GEIKIE, J. 1866. On the metamorphic Lower Silurian rocks of Carrick, Ayrshire. *Q. J. Geol. Soc. London*, Vol.20, 513–534.

GREIG, D. C. 1971. *British regional geology: the south of Scotland.* (Edinburgh: HMSO.)

HAMILTON, P. J., BLUCK, B. J., and HOLLIDAY, B. J. 1984. Sm-Nd ages from the Ballantrae complex, SW Scotland. *Trans. R. Soc. Edinburgh: Earth Sci.*, Vol.75, 183–187

HARLAND, W. B., COX, A. V., LLEWELLYN, P. G., PICKTON, C. A. G., SMITH, A. G., and WALTERS, R. 1982. *A geologic time scale* (2nd edition). (Cambridge and London: Cambridge University Press.)

HARRIS, P. M., FARRAR, E., MACINTYRE, R. M., and YORK, D. 1965. Potassium-argon age measurements on two igneous rocks from the Ordovician system of Scotland. *Nature, London,* Vol.205, 352–353.

HAWKINS, J. W., and EVANS, C. A. 1983. Geology of the Zambales Range, Luzon, Phillippine Islands: ophiolite derived from an island arc–back arc pair. 95–123 *in* The tectonic and geologic evolution of southeast Asian seas and islands: Part 2. HAYES, D. E. (editor). *Geophys. Monogr. Am. Geophys. Union*, No.27.

HEDDLE, M. F. 1878. Chapters on the mineralogy of Scotland, 4: augite, hornblende and serpentinous change. *Trans. R. Soc. Edinburgh*, Vol.28, 463–464.

HENDERSON, S. M. K. 1935. Ordovician submarine disturbances in the Girvan district. *Trans. R. Soc. Edinburgh*, Vol.58, 487–510.

HOLUB, F. V., KLAPOVA, H., BLUCK, B. J., and BOWES, D. R. 1984. Petrology and geochemistry of post-obduction dykes of the Ballantrae complex, SW Scotland. *Trans. R. Soc. Edinburgh: Earth Sci.*, Vol.75, 211–223.

HUBERT, J. F. 1966. Sedimentary history of Upper Ordovician geosynclinal rocks, Girvan, Scotland. *J. Sediment. Petrol.*, Vol.36, 677–699.

HUBERT, J. F. 1969. Late Ordovician sedimentation in the Caledonian geosyncline, southwestern Scotland. 267–283 *in* North Atlantic geology and continental drift: a symposium. KAY, M. (editor). *Mem. Am. Assoc. Petrol. Geol.*, No.12.

INCE, D. M. 1983. Shallow water facies and environments in the Ordovician of the Girvan district, Strathclyde. Unpublished PhD thesis, University of Edinburgh. 199pp.

INCE, D. M. 1984. Sedimentation and tectonism in the Middle Ordovician of the Girvan district, SW Scotland. *Trans. R. Soc. Edinburgh: Earth Sci.*, Vol. 75, 225–237.

INGHAM, J. K. 1978. Geology of a continental margin 2: middle and late Ordovician transgression, Girvan. 163–176 *in* Crustal evolution in northwestern Britain and adjacent regions. BOWES, D. R., and LEAKE, B. E. (editors). *Geol. J. Spec. Issue*, No. 10.

INSTITUTE OF GEOLOGICAL SCIENCES. 1980. 1:250 000 aeromagnetic anomaly map, Clyde sheet, 55°N–06°W. (Southampton: Ordnance Survey.)

JACKSON, E. D., GREEN, H. W., and MOORES, E. M. 1975. The Vourinos ophiolite, Greece: cyclic units of lineated cumulates overlying harzburgite tectonite. *Bull. Geol. Soc. Am*, Vol.86, 390–398.

JELINEK, E., SOUCEK, J., BLUCK, B. J., BOWES, D. R., and TRELOAR, P. J. 1980. Nature and significance of beerbachites in the Ballantrae ophiolite, SW Scotland. *Trans. R. Soc. Edinburgh: Earth Sci.*, Vol. 71, 159–180,

JELINEK, E., SOUCEK, J., RANDA, Z., JAKES, P., BLUCK, B. J., and BOWES, D. R. 1984. Geochemistry of peridotites, gabbros and trondhjemites of the Ballantrae complex, SW Scotland. *Trans. R. Soc. Edinburgh: Earth Sci.*, Vol.75, 193–209.

JONES, C. M. 1977. The Ballantrae complex as compared to the ophiolites of Newfoundland. Unpublished PhD thesis, University of Wales. 417 pp.

KELLING, G. 1961. The stratigraphy and structure of the Ordovician rocks of the Rhinns of Galloway. *Q. J. Geol. Soc. London.*, Vol.117, 37–75.

KUENEN, PH. H. 1953. Graded bedding, with observations on lower Palaeozoic rocks of Britain. *Ned. Akad. Wet., Afd. Natuurkd.*, Vol.20, No.3, 1–47.

LAMONT, A., and LINDSTRÖM, M. 1957. Arenigian and Llandeilian cherts identified in the Southern Uplands of Scotland by means of conodonts. *Trans. Geol. Soc. Edinburgh*, Vol.17, 60–70.

LAPWORTH, C. 1882. The Girvan succession. *Q. J. Geol. Soc. London*, Vol.38, 537–666.

LAPWORTH, C. 1889. On the Ballantrae rocks of South Scotland and their place in the Upland sequence. *Geol. Mag.*, Vol.6, 20–24 and 59–69.

LEGGETT, J. K., MCKERROW, W. S., and CASEY, D. M. 1982. The anatomy of a Lower Palaeozoic forearc: the Southern Uplands of Scotland. 494–520 *in* Trench-forearc geology: sedimentation and tectonics on modern and ancient active plate margins. LEGGETT, J. K. (editor). *Spec. Publ. Geol. Soc. London*, No.10.

Leggett, J. K., McKerrow, W. S., and Eales, M. H. 1979. The Southern Uplands of Scotland: a Lower Palaeozoic accretionary prism. *J. Geol. Soc. London*, Vol. 136, 755–770.

Leggett, J. K., McKerrow, W. S., and Soper, N. J. 1983. A model for the crustal evolution of southern Scotland. *Tectonics*, Vol.2, 187–210.

Lewis, A. D. 1975. The geochemistry and geology of the Girvan-Ballantrae ophiolite and related Ordovician volcanics in the Southern Uplands of Scotland. Unpublished PhD thesis University of Wales. 378 pp.

Lewis, A. D., and Bloxam, T. W. 1977. Petrotectonic environments of the Girvan-Ballantrae lavas from rare-earth element distributions. *Scott. J. Geol.*, Vol.13, 211–222.

Lewis, A. D., and Bloxam, T. W. 1980. Basaltic macadam-breccias in the Girvan-Ballantrae Complex, Ayrshire. *Scott. J. Geol.*, Vol.16, 181–188.

Longman, C. D., Bluck, B. J., and van Breemen, O. 1979. Ordovician conglomerates and the evolution of the Midland Valley. *Nature, London,* Vol.280, 578–581.

Lowe, D. R. 1982. Sediment gravity flows: II. Depositional models with special reference to the deposits of high-density turbidity currents. *J. Sediment. Petrol.*, Vol.52, 279–297.

Magaritz, M., and Taylor, H. P. 1974. Oxygen and hydrogen isotope studies of serpentinization in the Troodos ophiolite complex, Cyprus. *Earth & Planet. Sci. Lett.*, Vol.23, 8–14.

Malpas, J. 1979. Dynamothermal aureole beneath the Bay of Islands ophiolite in western Newfoundland. *Can. J. Earth Sci.*, Vol.16, 2086–2101.

McKerrow, W. S., Lambert, R. St. J., and Cocks, L. R. M. 1985. The Ordovician, Silurian and Devonian periods. 73–80 *in* Geochronology and the geological record. Snelling, N. J. (editor). *Mem. Geol. Soc. London*, No.10.

Mendum, J. R. 1968. Unconformities in the Ballantrae volcanic sequence. *Trans. Leeds Geol. Assoc.*, Vol.7, 261–264.

Murchison, R. I. 1851. The Silurian rocks of Scotland. *Q. J. Geol. Soc. London*, Vol.7, 137–178.

Mykura, W. 1965. The age of the lower part of the New Red Sandstone of south-west Scotland. *Scott. J. Geol.*, Vol.1, 9–18.

Nicol, J. 1844. *Guide to the geology of Scotland.* (Edinburgh: Oliver and Boyd.)

Nicolas, A., Bouchez, J. L., Boudier, F., and Mercier, J. C. 1971. Textures, structures and fabrics due to solid-state flow in some European lherzolites. *Tectonophysics*, Vol.12, 55–86.

Oliver, G. J. H., Smellie, J. L., Thomas, L. J., Casey, D. M., Kemp, A. E. S., Evans, L. J., Baldwin, J. R., and Hepworth, B. C. 1984. Early Palaeozoic metamorphic history of the Midland Valley, Southern Uplands-Longford Down massif and the Lake District, British Isles. *Trans. R. Soc. Edinburgh: Earth Sci.*, Vol.75, 245–258.

PEACH, B. N., and HORNE, J. 1899. The Silurian rocks of Britain, I: Scotland. *Mem. Geol. Surv. UK.*

PEARCE, J. A., and CANN, J. R. 1973. Tectonic setting of basic volcanic rocks determined using trace element analyses. *Earth & Planet. Sci. Lett.*, Vol.19, 290–300.

PIKE, J. E. N., and SCHWARZMAN, E. C. 1977. Classification of textures in ultramafic xenoliths. *J. Geol.*, Vol.85, 49–61.

POWELL, D. W. 1978. Geology of a continental margin 3: gravity and magnetic anomaly interpretation of the Girvan-Ballantrae district. 151–162 *in* Crustal evolution in northwestern Britain and adjacent regions. BOWES, D. R. and LEAKE, B. E. (editors). *Geol. J. Spec. Issue*, No.10.

RAMSAY, D. M., and STURT, B. A. 1973. An analysis of non-cylindrical and incongruous fold patterns from the Eocambrian rocks of Söröy, northern Norway. *Tectonophysics*, Vol.18, 81–107.

RICHEY, J. E. 1939. The dykes of Scotland. *Trans. Geol. Soc. Edinburgh*, Vol. 13, 393–435.

RICHEY, J. E., MACGREGOR, A. G., and ANDERSON, F. W. 1961. *British regional geology: Scotland: the Tertiary volcanic districts* (3rd edition). (Edinburgh: HMSO.)

ROBERTSON, A. H. F. 1982. Field-guide to aspects of the geology of the Ballantrae ophiolitic complex, SW Scotland. Compiled for a meeting of the Geological Society of London 'The geological evolution of the eastern Mediterranean'. (Unpublished.)

RUSHTON, A. W. A., STONE, P., SMELLIE, J. L., and TUNNICLIFF, S. P. 1986. An early Arenig age for the Pinbain sequence, Ballantrae Complex. *Scott. J. Geol.* Vol. 22, 41–54.

RUSHTON, A. W. A., and TRIPP, R. P. 1979. A fossiliferous lower Canadian (Tremadoc) boulder from the Benan Conglomerate of the Girvan district. *Scott. J. Geol.*, Vol.15, 321–328.

SEARLE, M. P., and MALPAS, J. 1980. Structure and metamorphism of rocks beneath the Semail ophiolite of Oman and their significance in ophiolite obduction. *Trans. R. Soc. Edinburgh: Earth Sci.*, Vol.71, 247–262.

SMELLIE, J. L. 1984a. Metamorphism of the Ballantrae Complex, south-west Scotland: a preliminary study. *Rep. Br. Geol. Surv.*, Vol.16, No.10, 13–17.

SMELLIE, J. L. 1984b. Accretionary lapilli and highly vesiculated pumice in the Ballantrae ophiolite complex: ash-fall products of subaerial eruptions. *Rep. Br. Geol. Surv.*, Vol.16, No.1, 36–40.

SMELLIE, J. L., and STONE, P. 1984. "Eclogite" in the Ballantrae Complex: a garnet-clinopyroxenite segregation in mantle harzburgite? *Scott. J. Geol.*, Vol. 20, 315–328.

SMITH, D. B., BRUNSTROM, R. G. W., MANNING, P. I., SIMPSON, S., and SHOTTON, F. W. 1974. A correlation of Permian rocks in the British Isles. *J. Geol. Soc. London*, Vol.130, 1–45.

Spray, J. G. 1982. Mafic segregations in ophiolite mantle sequences. *Nature, London*, Vol.299, 524–528.

Spray, J. G., and Williams, G. D. 1980. The sub-ophiolite metamorphic rocks of the Ballantrae Igneous Complex, SW Scotland. *J. Geol. Soc. London*, Vol.137, 359–368.

Stone, P. 1982. Clastic rocks within the Ballantrae complex: borehole evidence. *Rep. Inst. Geol. Sci.*, No.82/1, 45–47.

Stone, P. 1984. Constraints on genetic models for the Ballantrae complex, SW Scotland. *Trans. R. Soc. Edinburgh: Earth Sci.*, Vol.75, 189–191.

Stone, P., Floyd, J. D., Barnes, R. P., and Lintern, B. C. (*in press*). A sequential back-arc and foreland basin thrust duplex model for the Southern Uplands of Scotland. *J. Geol. Soc. London.*

Stone, P., Gunn, A. G., Coats, J. S., and Carruthers, R. M. 1986. Mineral exploration in the Ordovician Ballantrae Complex, SW Scotland. 265–278 in *Metallogeny of basic and ultrabasic rocks.* Gallagher, M. J. (editor). (London: Institution of Mining and Metallurgy).

Stone, P., Lambert, J. L. M., Carruthers, R. M., and Smellie, J. L. 1984. Concealed ultramafic bodies in the Ballantrae Complex, south-west Scotland: borehole results. *Rep. Br. Geol. Surv.*, Vol.16, No.1, 41–45.

Stone, P., and Rushton, A. W. A. 1983. Graptolite faunas from the Ballantrae ophiolite complex and their structural implications. *Scott. J. Geol.*, Vol.19, 297–310.

Stone, P., and Strachan, I. 1981. A fossiliferous borehole section within the Ballantrae ophiolite. *Nature, London*, Vol.293, 455–456.

Thirlwall, M. F., and Bluck, B. J. 1984. Sr-Nd isotope and geological evidence that the Ballantrae ''ophiolite'', SW Scotland, is polygenetic. 215–230 *in* Ophiolites and oceanic lithosphere. Gass, I. G., Lippard, S. J., and Shelton, A. W. (editors). *Spec. Pub. Geol. Soc. London*, No. 13.

Treloar, P. J., Bluck, B. J., Bowes, D. R., and Dudek, A. 1980. Hornblende-garnet metapyroxenite beneath serpentinite in the Ballantrae complex of SW Scotland and its bearing on the depth provenance of obducted oceanic lithosphere. *Trans. R. Soc. Edinburgh: Earth Sci.*, Vol.71, 201–212.

Tripp, R. P. 1980. Trilobites from the Ordovician Balclatchie and lower Ardwell groups of the Girvan district. *Trans. R. Soc. Edinburgh: Earth Sci.*, Vol.71, 123–145.

Tripp, R. P., Williams, A., and Paul, C. R. C. 1981. On an exposure of the Ordovician *superstes* Mudstones at Colmonell, Girvan district, Strathclyde. *Scott. J. Geol.*, Vol.17, 21–25.

Tyrrell, G. W. 1909. A new occurrence of picrite in the Ballantrae District and its associated rocks. *Trans. Geol. Soc. Glasgow*, Vol.13, 283–290.

Walton, E. K. 1961. Some aspects of the succession and structure in the Lower Palaeozoic rocks of the Southern Uplands of Scotland. *Geol. Rundsch.*, Vol.50, 63–77.

WALTON, E. K. 1983. Lower Palaeozoic—structure and palaeogeography. 139–166 in *Geology of Scotland* (2nd edition). CRAIG, G. Y. (editor). (Edinburgh: Scottish Academic Press.)

WENNER, D. B., and TAYLOR, H. P. 1971. Temperatures of serpentinization of ultramafic rocks based on O^{18}/O^{16} fractionation between coexisting serpentine and magnetite. *Contrib. Mineral. Petrol.*, Vol.32, 165–185.

WENNER, D. B., and TAYLOR, H. P. 1974. D/H and O^{18}/O^{16} studies of serpentinization of ultramafic rocks. *Geochim. Cosmochim. Acta*, Vol.38, 1255–1286.

WICKS, F. J., and WHITTAKER, E. J. W. 1977. Serpentine textures and serpentinization. *Canadian Mineral.*, Vol.15, 459–488.

WICKS, F. J., WHITTAKER, E. J. W., and ZUSSMAN, J. 1977. An idealized model for serpentine textures after olivine. *Canadian Mineral.*, Vol.15, 446–458.

WILKINSON, J. M., and CANN, J. R. 1974. Trace elements and tectonic relationships of basaltic rocks in the Ballantrae igneous complex, Ayrshire. *Geol. Mag.*, Vol.111, 35–41.

WILLIAMS, A. 1959. A structural history of the Girvan district of south-west Ayrshire. *Trans. R. Soc. Edinburgh*, Vol.63, 629–667.

WILLIAMS, A. 1962. The Barr and Lower Ardmillan Series (Caradoc) of the Girvan district, south-western Ayrshire, with descriptions of the brachiopoda. *Mem. Geol. Soc. London*, No.3, 267 pp.

WILLIAMS. G. D., and SPRAY, J. G. 1979. Non-cylindrical, flexural slip folding in the Ardwell Flags—a statistical approach. *Tectonophysics*, Vol.58, 269–277.

WOODCOCK, N. H., and ROBERTSON, A. H. F. 1982. Wrench and thrust tectonics along a Mesozoic-Cenozoic continental margin: Antalya Complex, SW Turkey. *J. Geol. Soc. London*, Vol.139, 147–163

Excursion itineraries

Most of the main elements of the Ballantrae Complex, together with the overlying Ardmillan Group and the neighbouring Downan Point Lava Formation, can be conveniently examined in coastal exposures adjacent to the A77 (T) road. Localities 1 – 10 below are particularly recommended as readily accessible areas in which much of interest is exposed. Three important aspects of the Ballantrae Complex which cannot be observed on the coast are the trondhjemite intrusion into the northern serpentinite belt, the metamorphic aureole at the south-east margin of the same ultramafic body, and the beerbachite xenoliths in the southern serpentinite belt. These can be examined at inland exposures (localities 11 – 13), all of which are on privately owned farmland. Prior permission is, therefore, required if they are to be visited. Further details of all localities may be found in the main body of the preceding text or in the references cited. For convenience the localities are numbered on the back end paper map.

1 Kennedy's Pass [NX 1492 9327]
Parking is available in a large lay-by on the seaward side of the road. The Kilranny Conglomerate member of the Balclatchie Formation is here exposed, unconformably overlain by greywacke of the Ardwell Formation (both are Caradocian parts of the Ardmillan Group). Northwards along the coast for 500 – 600 m the turbiditic greywackes and siltstones of the Ardwell Formation have been deformed by a spectacular development of cascading box folds (Figure 30; cf. Williams and Spray, 1979).

2 Slockenray [NX 1408 9198] Parking for several cars is available in a lay-by on the seaward side of the road. The local sequence of aphyric and feldsparphyric lavas, breccias and hyaloclastites has been described in detail by Bluck (1982) and interpreted as a shallow-water, deltaic succession. It forms part of the Slockenray Formation, a component of the northern outcrop of the Arenig Balcreuchan Group.

3 Pinbain Bridge [NX 1374 9141]
Parking for several cars is available in a lay-by on the seaward side of the road. Arenig mélange is here faulted against serpentinite with Tertiary dykes intruded into the contact zones. The mélange contains a wide variety of clasts (Figure 14; cf. Bailey and McCallien, 1957; Bluck, 1978) with a unit of brecciated pillow lava in its central zone which may be a very large clast or a discrete horizon. At its northern margin the mélange and breccia are separated by a thin sliver of faulted serpentinite from the main northern outcrop of the Balcreuchan Group, the southernmost beds of which contain a sparse graptolite fauna of early Arenig age (Rushton and others, 1986). The most fossiliferous exposures are in the cliff face on the landward side of the road immediately to the north of the small raised-beach embayment formed by the preferential weathering of the faulted serpentinite sliver. In the upper part of the back wall of this embayment serpentinite and layered dunite-chromitite are exposed (Stone and others, 1986). A walk of about 400 m southwards along the beach is required to reach the next locality, Bonney's Dyke.

4 Bonney's Dyke [NX 1353 9106] is the colloquial name for a striking pegmatitic gabbro contained within the ultramafic northern serpentinite belt. The gabbro is medium to very coarse grained, formed of rodingitised clinopyroxene and plagioclase, with marked and random internal variations in texture (Figure 6). The ultramafic rock adjacent to the pegmatite is principally serpentinised harzburgite cut by a network of coarse clinopyroxenite veins.

The northern margin of the gabbro has been sheared and extensively rodingitised.

5 Carleton Fishery [NX 1260 8945] Ample parking is available in the large picnic area on the seaward side of the road 200 m west of the outcrop. The rocky sea stacks beside the boathouse are dolerite intrusions into the northern serpentinite belt, members of the late Arenig dyke swarm. Their margins are chilled against the serpentinite and are intensely rodingitised. Most of the ultramafic rock in the vicinity is serpentinised harzburgite but some fine compositional layering between the harzburgite, dunite and pyroxenite is developed locally.

6 Games Loup [NX1048 8807] Parking is available above the outcrop at Troax Bridge in lay-bys on either side of the road; thence a steep scramble is required to the shoreline. The southern marginal fault of the northern serpentinite is there exposed. Sheared and brecciated lavas to the south-east are juxtaposed against serpentinised harzburgite to the north which shows two distinct colour variations. At the eastern end of the beach section the serpentinite is dark green with paler irregular segregations of pyroxenite. Westward there is an apparent interfingering of the green serpentinite with a red variety until, where the red serpentinite is dominant, the green serpentinite appears as 'veins' cutting the red variety (Figure 5). The only significant difference in the two colour variants is the higher hematite content of the red serpentinite. South-east of the fault the Games Loup lava sequence consists of well formed, basaltic pillows with very little interbedded or interpillow sedimentary rock.

7 Balcreuchan Port [NX 0995 8752] Ample parking is available in the large lay-by overlooking this spectacular, precipitous cove. A footpath steeply descends the cliffs on its northern side: take great care. All the surrounding crags expose pillow basalt and breccia, the southern part of the Games Loup sequence. Approximately north–south faulting defines the eastern side of the bay where, below high-water mark, sheared and carbonatised serpentinite is exposed, faulted against the lavas. A prominent Tertiary basalt dyke is intruded into the marginal serpentinite. The traverse for about 500 m south-west along the coast, a fairly easy scramble at low tide but difficult at high water, crosses part of the Balcreuchan Group succession (Fig. 16). Brecciated lavas, still part of the Games Loup sequence, are dominant as far as a major north-south fault zone. Across this fault the lithology becomes much more variable with aphyric and feldsparphyric lavas alternating with chert, sandstone and breccia (Table 5). The sedimentary rocks contain a sparse early Arenig graptolite fauna and the whole sequence has been structurally imbricated (Stone and Rushton, 1983).

8 Bennane Head [NX 0913 8667] To examine roadside exposures of macadam breccia it is best to park in the large lay-by above Port Vad and walk about 300 m south-west along the road. The breccias are well exposed in both the sea cliffs and the cutting face on the inland side of the road, where they are interbedded with graded units of basaltic sandstone. They form the central part of the Balcreuchan Group's Bennane Head outcrop (Table 5).

9 Bennane Lea [NX 0918 8605] Parking for several cars is available in the lay-by on the seaward side of the road adjacent to the beach. In the intertidal zone to the south, Permian red sandstone crops out and at low tide can be seen to unconformably overlie ultramafic rocks of the Arenig Ballantrae Complex (Fig. 16). The ultramafic serpentinites, together with numerous gabbro enclaves, are exposed in a narrow zone between the unconformity and the faulted, southern margin of the Balcreuchan Group; here represented by interbedded chert and conglomerate (Table 5; cf. Bluck, 1978). The succession is much confused by slumping and tectonism, the latter producing a series of tight folds plunging seawards and well exposed in the old sea cliffs on the inland side of the raised beach. Northwards along the coast, generally down sequence, the bedded cherts are underlain by a zone in which basaltic sandstone, breccia and mudstone alternate with chert. This assemblage is then underlain by massive basalt flows between which black, siliceous mudstone is sporadically developed. At one locality middle Arenig graptolites have been collected (Table 5; cf. Stone and Rushton, 1983).

10 Sgavoch Rock [NX 0745 8098] About 2 km south-south-west of Ballantrae, on the coast opposite the Sgavoch Rock and just beyond the southern margin of the 1:25 000 map of the Ballantrae area, one of the most impressive series of pillow lavas to be seen in Britain is exposed (Figure 25; cf. Bloxam, 1960). Access is via a series of minor roads leading west from the A77 immediately south of Ballantrae and the River Stinchar. Cars may be taken to the bottom of a track (NX 0792 8133) which leaves the Downan Farm road about 1 km before the farm is reached. The basalt pillows and rare intercalated chert form part of the Downan Point Lava Formation. Individual pillows range from elliptical, through elongate bolster-shape, to sheet flows. Most pillows have a highly vesicular crust, the larger vesicles tending to occur in concentric bands parallel to the upper margin. In some pillows partial drain-back has left spaces which are now infilled with carbonate.

11 Byne Hill [NX 1815 9450] Trondhjemite, its transition to gabbro and its chilled relationship with the enclosing serpentinite are best seen on the eastern slopes of Byne Hill (Bloxam, 1968). The exposures may be reached via a footpath from Brochneil (NX 1875 9534) just beyond the northern margin of the Ballantrae 1:25 000 sheet.

12 Water of Lendal The metamorphic aureole at the south-east margin of the northern serpentinite belt is fragmentary and only poorly exposed. None of the localitites available is really suitable for examination by a large party. The higher-grade parts of the aureole, amphibolite and garnet metapyroxenite (Treloar and others, 1980) can be seen in and around streams draining from the north-west into the Water of Lendal near Knocklaugh (NX 1685 9193). The lower-grade epidote and chlorite schists are best exposed in the Water of Lendal just upstream from Straid Bridge (NX 1394 9006) and on the nearby slopes of Carleton Hill (NX 1303 8945).

13 Garna Burn [NX 1572 8745] Beerbachite xenoliths in the southern serpentinite belt can be most readily examined in the Garna Burn tributary stream section and around its headwaters on the western side of Craig Hill (NX 1625 8765). The enclosing ultramafic rock is predominantly serpentinised harzburgite. The Garna Burn area is best reached from the minor road which links Colmonell and Lendalfoot.

Glossary

For additional or alternative definitions to those given below see The Penguin Dictionary of Geology (D. G. A. Whitten and J. R. V. Brooks) or the Americal Geological Institute Glossary Geology (R. L. Bates and J. A. Jackson).

Abyssal Relating to rocks deposited at oceanic depths in excess of about 2000 m or otherwise formed in a deep oceanic environment.

Accretionary prism A complex of inclined strata formed above an active subduction zone by the sequential stacking, through underthrusting, of successively younger units of oceanic sedimentary and volcanic rocks.

Acidic Used to describe igneous rocks with more than about 10% free quartz or with a total silica content in excess of 66%.

Allotriomorphic A textural term applied to igneous rocks in which the development of crystal form is prevented by the crowding together of adjacent mineral grains during crystallisation. It is a special case within the general 'anhedral' texture.

Anhedral A term applied to describe the texture of an igneous rock in which the mineral grains show no development of crystal form.

Aphyric A textural term used to describe a uniformly fine-grained igneous rock in which there are no phenocrysts.

Augen Elongate clusters of coarse crystals which develop parallel to the foliation of metamorphic rocks.

Autoclastic Having a broken or brecciated condition imposed on the rock during its original formation and not as a result of subsequent geological processes.

Basement Igneous and metamorphic rocks forming the stable, deeper levels of the continental crust. It may be exposed through long-term uplift and erosion.

Basic A term describing an igneous rock which contains no free quartz, contains feldspars which are generally more calcic than sodic, or which has less than about 50% total silica.

Beerbachite A thermally metamorphosed rock with a well developed, polygonal-granoblastic texture developed from an original dolerite or gabbro.

Biosparite A limestone consisting of broken shell fragments and fragmented or complete calcareous organisms such as bryozoa, forams or algal debris, all cemented together by comparitively clear, recrystallised calcite.

Blueschist A metamorphic rock formed under high-pressure/low-temperature conditions having a schistose fabric and containing amphiboles such as glaucophane and crossite. These give the rock a characteristic blue colour.

Boudins The result of tensional stress in a bed or layer of rock. Deformation is initially by necking or pull-apart to produce eventually a structure which, in cross section, looks like a string of sausages (French *boudin*, sausage).

Cataclastic Having a fractured and brecciated fabric caused by faulting or dynamic metamorphism. The grain size of the rock has generally been reduced by mechanical crushing and milling of the original components.

Connate water Water trapped in sediments at the time of deposition.

Convolute bedding A sedimentary structure in which bedding laminae are contorted into a series of markedly irregular folds, the distortion increasing upwards, which may be abruptly truncated by overlying strata. The folding, which is confined to one bed, may be produced by post-depositional dewatering or slumping.

Crenulation cleavage Cleavage planes, whether micaceous layers or sharp breaks, which are separated by thin slices of rock containing a marked cross lamination.

Cumulate An igneous rock formed by the accumulation of crystals which precipitate from a magma before any modification by later crystallisation.

Cumulophyric A variation of porphyritic texture in which clusters of crystals are scattered through a finer-grained ground mass.

Diagenetic Processes affecting a sediment, whilst it is at or near the Earth's surface, during its lithification.

Dynamothermal The combined metamorphic effect of heat and movement.

Euhedral A term applied to mineral grains displaying fully developed crystal form.

Exsolve The separation of an initially homogeneous solid solution into two or more distinct crystalline phases with no change in the bulk composition. The process of exsolution generally occurs on cooling.

Feldsparphyric A specific case of porphyritic texture in which the phenocrysts are exclusively of plagioclase.

Felsic A collective term for light-coloured minerals such as quartz and feldspar and for rocks composed mainly of those minerals.

Flaser A streaky structure developed in granular igneous rocks by dynamic metamorphism.

Granoblastic A texture in metamorphic rock in which recrystallisation has formed equidimensional polygonal crystals.

Granulite facies The set of metamorphic mineral assemblages caused by deep-seated regional or dynamothermal metamorphism at temperatures in excess of about 650°C. Under these conditions basic rocks are represented by diopside + hypersthene + plagioclase.

Graphic An igneous texture in which an intergrowth of quartz and alkali feldspar produces a distinctive pattern reminiscent of cuneiform or runic writing.

Greenschist facies The set of metamorphic mineral assemblages caused by low-grade regional metamorphism in the temperature range 300 – 500°C. Under these conditions basic rocks are represented by albite + epidote + chlorite + actinolite.

Heteroblastic A textural term applied to metamorphic rocks in which the relative crystal sizes of the essential constituents are of two or more distinctive orders of magnitude.

Hornfels A medium- or fine-grained granulose rock produced by thermal metamorphism but often retaining a relict premetamorphic fabric.

Hot spot An intraplate oceanic volcanic centre thought to be the surface expression of a persistent rising plume of hot mantle material: the likely cause of 'Hawaiian-type' oceanic islands.

Hyaloclastite A deposit formed by the flowing or intrusion of lava into water or wet sediment and its consequent shattering into small glassy fragments.

Imbrication 1. The structural repetition of strata by a series of subparallel thrust faults. 2. The current-imposed orientation of pebbles in a conglomerate whereby the long axes of the clasts lie parallel to each other 'leaning' in the direction of current flow.

Intercumulus Crystalline material occupying the interstices between the precipitated crystals in a cumulate igneous rock.

Intraplate A process or environment occurring within a geological plate rather than at its margins.

Island arc A chain of volcanic islands formed above a subduction zone, usually in an oceanic setting and adjacent to a deep ocean trench the surface expression of the subduction zone.

Kink band The long narrow zone defined by the short limb of a markedly assymmetric, open or close, small-scale fold pair which has planar limbs and very angular hinges.

Knocker A hard body of rock contained within a softer lithology which, through preferential erosion of the surrounding softer material, forms a prominent monolith.

Load casts Bulbous projections on the base of some sandstone beds caused by the variable response of the bed below to unequal loading. They are most common where a thick bed of coarse sandstone is underlain by a much finer-grained lithology.

Mafic A collective term for dark-coloured ferromagnesian minerals and for rocks composed mainly of those minerals.

Mantle That portion of the Earth's interior below the Moho (a depth of about 35 km) down to about 2900 km, and consisting mainly of olivine.

Marginal basin A basin formed by limited generation of ocean crust between a volcanic island arc and a continental margin.

Mélange A heterogeneous rock type in which a variety of exotic rock clasts are contained in a pervasively foliated, fine-grained matrix. The matrix may be a sedimentary mudstone or a serpentinite.

Metasomatism A form of metamorphism involving the introduction of material from an external source.

Meteoric water Water that penetrates the rocks from above and originates at the Earth's surface.

Mylonite A fine-grained, usually foliated, cataclastic rock produced by the shearing and dislocation effects of dynamic metamorphism.

Myrmekitic An igneous texture in which an intergrowth of quartz and plagioclase feldspar produces a distinctive pattern of interlocking worm-like rods.

Nematoblastic A term applied to metamorphic rocks in which a foliated texture is due to the development during recrystallisation of abundant, slender prismatic crystals.

Neoblast A crystal of new mineral formed by the metamorphic recrystallisation of a usually finer-grained matrix.

Neritic Relating to rocks deposited on, or the environment existing on, a continental shelf beyond the intertidal zone.

Obduction The large-scale thrusting of oceanic crust onto a continental margin.

Oceanic crust The assemblage of extrusive and intrusive igneous rocks generated at constructive plate margins mostly along the mid-ocean ridges.

Oceanic island An intraplate volcanic edifice formed in an oceanic setting above a hot spot.

Oligomict Applied to breccias and conglomerates which consist of fragments of only one rock type.

Olistostrome A massive and chaotic conglomerate which accumulated in deep water as a result of large-scale slumping.

Ophicarbonate A rock consisting mainly or entirely of secondary quartz and carbonate, both dolomitic and calcitic, formed by the alteration of serpentinite.

Ophiolite An assemblage of petrogenetically related igneous rocks which form a 'pseudostratigraphic' upward sequence of ultramafic rocks, layered and static gabbros, sheeted dykes and extrusive basalt lavas. This series has a great similarity to the rock associations in oceanic crust, and many ophiolites probably originated at active mid-ocean spreading ridges. However, other origins are possible and the term carries no genetic implications.

Ophitic A texture characteristic of basic igneous rocks in which large augite crystals enclose, either partially or wholly, laths of plagioclase.

Overfold A tight or isoclinal fold which is overturned so that both limbs and the axial plane all dip in the same direction.

Pericline A basin or dome structure in which the dip of the folded strata is 'quaquaversal', i.e., it either converges towards, or radiates from, a single point.

Phenoclast A general term for a fragment larger than about 4 mm diameter contained in a sedimentary rock, i.e. a pebble, cobble or boulder.

Phenocryst One of the relatively large, usually euhedral crystals which, when set in a fine-grained groundmass, produce a porphyritic igneous texture.

Plunge A fold is said to plunge if the axis is not horizontal, the amount of plunge being the angle between the axis and the horizontal line lying in the same vertical plane.

Poikilitic A texture in an igneous rock caused by the total enclosure of a number of small crystals of one mineral in a single, large crystal of a second mineral.

Porphyrite A fine-grained, intrusive igneous rock with phenocrysts of albitised plagioclase, hornblende, pyroxene and/or biotite all contained within a highly altered groundmass.

Porphyritic A texture in igneous rocks caused by the presence of relatively large crystals contained in a markedly finer-grained groundmass. On a macroscopic scale this may give the rock a spotted appearance. When the same phenomenon occurs on a microscopic scale *micro-porphyritic* is used.

Porphyroclast A relict crystal or fragment of crystal in a metamorphic rock which is contained in a markedly finer-grained crushed or recrystallised matrix.

Prehnite-pumpellyite facies The set of metamorphic mineral assemblages caused by pressure-temperature conditions intermediate between those of the zeolite and greenschist facies. It is regional, low grade, and characterised in some basic rocks by the association albite + quartz + prehnite + pumpellyite + chlorite + sphene.

Protolith An original lithology which has been converted by metamorphism or metasomatism into a different, secondary rock type.

Pseudomorph A secondary mineral occurring in the crystal form of a different, primary mineral usually by replacement through alteration.

Pyroclastic Consisting of fragmental volcanic material which has been blown into the atmosphere by explosive activity and subsequently fallen back to the Earth's surface.

Reclined A descriptive term for a fold which closes sideways such that the fold axis plunges down dip on the axial plane.

Reticulate An intricate, net-like mesh pattern, usually applied to vein complexes.

Retrogressive metamorphism The process whereby metamorphic minerals of a lower grade are formed at the expense of minerals characteristic of higher grades during readjustment to a reduction in temperature and/or pressure.

Rodingite A doleritic rock which has suffered extensive calcium metasomatism and now consists almost exclusively of calcium-rich secondary minerals such as prehnite, pectolite and grossular.

Sheeted dykes A complex of multiple dykes with a very few or no intervening screens of country rock. Ideally the complex is on a scale of at least several hundred metres with individual dykes showing a consistent sense of chilling away from a central point. It is believed to result from repeated intrusion at a mid-ocean ridge during sea floor spreading.

Slaty cleavage A penetrative fabric caused by mineral orientation in certain favoured layers in a rock. These layers are generally very close-spaced.

Slickenside Mechanical striation or grooving of a fault plane or other surface (e.g. bedding) along which there has been movement.

Spreading ridge The active volcanic zone in mid-ocean at which new oceanic crust is being generated and from which two plates are diverging.

Strike-slip Used to describe the movement on a vertical or subvertical fault where the net slip direction is near horizontal.

Stromatolite A variable layered structure caused by the trapping and binding of sediment during the growth of blue-green algae.

Subduction trench The surface expression of the contact line between two converging plates where one, usually oceanic, plunges beneath the other and is ultimately destroyed. Such trenches are the deepest parts of the oceans and are often associated with island arcs.

Subsolidus recrystallisation Recrystallisation reactions occurring in the solid state at temperatures below the melting point of the original mineral.

Tectonic inclusion An exotic block of rock introduced into a markedly different host lithology either by faulting or, in the case of a serpentinite host, by low-temperature flow of the solid rock.

Thermal aureole The zone of metamorphosed rock surrounding an intrusive mass of igneous rock.

Thin-skinned A descriptive term for tectonic processes occurring at relatively small crustal depths above a major basal fault or thrust.

Transform fault A large-scale structure along which there has been considerable lateral movement either at the boundary of a plate or cutting across and displacing its margin.

Turbidite The lithified sediment originally deposited from a turbidity current. Typically cyclic units are seen grading upwards from conglomerate or coarse greywacke at the base, through parallel and cross-laminated siltstone, to mudstone or shale at the top.

Ultramafic A descriptive term applied to igneous rocks consisting almost exclusively of ferromagnesian minerals such as olivine and pyroxene.

Variolitic A textural term applied to devitrified glassy basic lavas which have developed spherules of radiating fibres, generally of plagioclase.

Vesicle A small spherical cavity in a volcanic lava produced by bubbles of gas trapped during solidification of the rock.

Volcaniclastic A descriptive term for a sedimentary rock composed mainly or exclusively of volcanic debris.

Xenolith An inclusion of pre-existing rock contained within an igneous rock.

Younging In an inclined sedimentary sequence the younging direction is that direction, perpendicular to strike, in which successively younger strata crop out.

Zeolite facies The set of metamorphic mineral assemblages caused by the lowest grade of regional metamorphism at low pressures and temperatures below about 250–300°C. At the lower end the facies is transitional to diagenesis.

Index of geographical localities